# Rules for Leadership

# Rules for Leadership
### IMPROVING UNIT PERFORMANCE

## Jon W. Blades

### WITH AN INTRODUCTION BY
### WALTER F. ULMER, JR.

**Fredonia Books**
**Amsterdam, The Netherlands**

Rules for Leadership:
Improving Unit Performance

by
Jon W. Blades

ISBN: 1-4101-0864-3

Reprinted from the 1986 edition

Fredonia Books
Amsterdam, The Netherlands
http://www.fredoniabooks.com

**DUTY * HONOR * COUNTRY**

*This book is dedicated to
leaders of all ranks who aspire to
these values*

J.W.B.

# Contents

# Rules

# Tables and Figures

**Tables**

## Figures

# Foreword

What is successful leadership and why are some leaders more successful than others? Searching for the answer, Lieutenant Colonel Jon Blades, US Army—an experienced troop commander and scholar of leadership theory—has studied the performance of Army leaders and units for most of his career. This book reflects both his experience and extensive research.

Literature on the subject of leadership contains divergent conclusions about how style, intelligence, ability, motivation, cohesion, and standards affect unit performance. Colonel Blades confirms the value of each of these elements and then explains the relationships among them. Most important, he presents practical rules leaders might apply in varying situations to improve the mission performance of their units.

It is appropriate that Colonel Blades wrote his findings during 1985, the Army's "Year of Leadership." The National Defense University is pleased to publish this contribution to the *art of leadership*.

Richard D. Lawrence
Lieutenant General, US Army
President, National Defense University

# Preface

This book presents an original set of leadership "rules" or principles that can be used to improve unit performance in any specific group situation and at all organizational levels. The 10 "rules" describe the influence which leadership style, leader enforcement of performance standards, group member intelligence, group member ability, leader intelligence, leader ability, group member motivation, leader motivation, and group cohesion have on unit performance.

Basically, the evidence establishes very clearly that (1) the nondirective and directive leadership styles are each very effective ways to improve unit performance in certain situations but are poor choices to use in others; (2) high levels of enforcement of performance standards, intelligence, ability, motivation, and cohesion will increase unit performance in some situations but will not influence the outcome in others; and (3) low levels of these factors will decrease unit performance in certain situations but will not affect the outcome in others. In other words, the effect each of these factors has on performance is not constant but, rather, varies from one situation to the next. The key premise is that the amount of influence each of these factors has on performance depends upon the presence or absence of certain other factors in the group situation. For each rule in this book, statistical evidence in the form of data collected from 49 US Army units is presented in support. The summary chapter integrates the separate rules and presents recommended courses of action for the leader to take in order to use the rules properly in whatever particular group situation is encountered.

In a secondary role, additional material is presented which describes actions that leaders can take to raise the level of several important group qualities such as motivation, ability, and cohesion.

# Acknowledgments

I wish to gratefully recognize Emile "Chum" Robert, Walter Ulmer, and Albert Akers for very generously providing assistance to me in writing this book. Their reviews of the draft manuscript, contributions, and comments were especially valuable.

Other friends and professional associates also assisted with suggestions and were very encouraging with their comments. John Johns, Norm Grunstad, Bob Haffa, and Steve Blades deserve special acknowledgment. Further, Kent Esbenshade's extensive editorial contribution is worthy of special mention.

Several individuals within the National Defense University Research and Publications Directorate should also be recognized for their assistance. I appreciate the work of Fred Kiley, Jan Hietala, and Pat Williams.

J.W.B.

# Introduction by
# Walter F. Ulmer, Jr.

There appears to be within America a resurgence of interest in the study and application of "leadership." Stung by the success of foreign companies in the US marketplace, aware that up-to-date technology is essential but yet not the tonic that ensures excellent corporate results, and excited by stories of organizational success stemming from executive vision and competence, Americans in many sectors have rediscovered the importance of leadership. While the military's enthusiasm for the subject has never really waned, its approach to leadership has wandered, been diffused and refocused, exuded confidence in or scorn for useful behavioral models, and debated the "born" versus "learned" theses.

Among nearly all students of the process and practice of leadership in recent years, the importance of varying the style to fit the specific situation has been recognized. Descriptors of differing "situations" have included maturity of the group, echelon within the organization, constraints of time and resources, effectiveness of the organization, and background of leaders and followers. But somehow we have missed the mark in developing a useful model which translates differing situational variables into a reasonably reliable prescription for leader action. One of the main reasons for the erratic responses and frequent disillusionment with leadership

Lieutenant General (Retired) Walter F. Ulmer, Jr., US Army, is the President, The Center for Creative Leadership, Greensboro, North Carolina.

models may be their relative disregard for two key ingredients in the leadership effectiveness formula: the skill and the motivation of both the leaders and followers. While situational settings are often described with care (the conference room or the tank gunnery range); and some leader attributes are routinely noted (age, experience, education); and the values and culture of the organization are mentioned as contributing to expectations and standards, we commonly find a lack of attention paid to group skill and motivation characteristics. These factors are assumed to be there or, perhaps, are considered as being in the "too hard" box. We may be a bit fearful of dissecting and revealing intellect and motivation. Surely in a perfect organization, intellect and motivation could be taken for granted—the differences among the participants being insignificant to the outcome. In reality, differences in intellect and motivation are gigantic factors, their importance growing with the complexity of the battlefield and the diversity of background and values which new soldiers—enlisted and commissioned—bring with them into the services.

Arguments in support of the absolute essentiality for some minimum level of skill (a combination of aptitude or intellect plus training) and a high motivation to perform with excellence toward achieving the unit mission need to be available and ready for use. For in the next decade we will see a public debate on the need for soldier quality whose outcome will set the limits of combat power and of deterrent power of our military formations for years to come. As our pool of eligible manpower diminishes amid an international environment which calls for more, not for less, active and reserve personnel, the issue of soldier quality will move to the forefront.

Significant contributions to behavioral science seem to be self-evident once somebody else has explained them clearly. Although other presentations have addressed the issues of leader and follower competence and motivation, Jon Blades has derived a set of rules which will make good sense to our experienced soldiers. This book is not a comprehensive tour across the fields of contemporary theory. Rather, it is an explanation of how several important factors affect unit performance. The author's intention is to provide common sense "rules" for practical application by unit leaders. Depending on one's background and expectations, the brief statistical excursions and the personal anecdotes and examples could be either comforting or not. Taken, however, in context, these elements make sense and provide a particular vitality to the "rules" which are the outcome. The material presented has enormous potential for explaining some heretofore neglected or unresolved gaps between experienced intuition and theoretical prediction.

This text, valuable to the junior leader and the researcher alike, should prompt discussions among practitioners of the art of leadership. After recognizing in the examples, rules, and outcomes the powerful synergistic effects of a combination of high leader and follower motivation and ability, we must recognize that the next crucial step must be some convenient method of measuring these components so that the leader can apply the correct approach. And, regularly, someone is going to have to measure the leader's motivation and ability. The fact that some percentage—even if small—of our leaders have low ability or motivation or both speaks loudly about our assessment and screening mechanisms. Understanding, then, the imperfection of our predictive or screening abilities combined with the impact of differing levels of ability and motivation, we need to develop practical methods for unobtrusively measuring these qualities in the field.

Jon Blades' discussion also raises useful issues regarding "cohesion," that marvelous word that too often connotes the ultimate objective instead of the powerful intermediate step which it is. As the book points out, cohesion alone is not enough. Cohesion alone won't guarantee success. As an aside, we might also notice that the crucial interpersonal bonds and shared experiences which are the elements of cohesive groups are by definition only possible within small formations. Probably company level is the maximum size. Talk of "cohesive brigades" is essentially nonsensical, although brigades could surely be comprised of company and platoon-sized units which are highly cohesive.

The discussion of standard setting and standard enforcing is clearly relevant to any presentation on effective unit leadership. The "standards" discourse in this book is worthy of a much lengthier treatment in some other related study. The problem is to determine which standards are the critical lynchpins between leadership and unit effectiveness. The rigid enforcement of meaningless or trivial "standards" is the hallmark of a disjointed organization. Sloppy standard setting, measuring, and enforcing can compromise the leader's reputation for skill, decrease follower motivation, and negate the great potential of follower intelligence. Our service schools have not done well in discussing standard setting, as they have not done well in teaching a coherent, functional method of measuring efficiency or effectiveness in field units.

After reading Jon Blades' work, a practicing leader must feel both helped and challenged. While these rules are utilitarian, they are, owing to the interactive nature of leadership, necessarily somewhat complex. The disconcerting fact is that the variables are not only difficult to measure quickly and accurately (motivation being a good example), but that the modification of one variable in the equation might likely effect the other variables as well as the outcome directly. As the leader or members improve their skills, for example, the appropriate style for the leader to use might be changed in the process. Thus the application of these rules to actual settings is by no means a simple task. Leadership is clearly as much an art as a science. Recognizing this, Jon Blades' derived rules should assist measurably by providing a solid basis and rationale for the "gut feel." But our work is far from over. And the ultimate stakes in this business remain enormously high.

# Rules for Leadership

# 1
# An Overview:
# Rules for Leadership

Although the study and practice of leadership are centuries old and, certainly, every generation has had a good number of excellent scholars and leaders, a single, widely accepted "theory of leadership" still does not exist. Among both academic theorists and military leaders, the two groups that produce the great majority of work and thought in this field, there continues to be little consensus about what factors cause group performance to be better or worse. Perhaps the only point upon which these people agree is that leadership plays a critical role in how well an organization performs. Whether we are talking about military commanders, elected officials, or business executives, good leaders are thought to cause more successful organizations and outcomes, while poor leaders are thought to produce less successful results.

### Background

During the past 35–40 years, theorists, primarily academicians, have done a great deal of work on research studies and leadership models. Although the research studies are a valuable source of ideas and supporting evidence for developing leadership theory, a reader notices two distinct problems when reviewing this work. First, the research contains several studies, some done by well-recognized authors, whose findings clearly go against what one intuitively feels is correct or has experienced while in a leadership position. For example, several studies report that group member intelligence appears to influence unit performance very little. Yet, which of us, when given the choice between having bright people or relatively slow people work for us, would not instantly opt for the bright people? Our intuition and experience tell us that the bright people, all other things being equal, will do a better job. Second, for each of the leadership factors this book discusses, the research contains several studies in which findings disagree considerably. For example, although several studies indicate that group member intelligence influences unit performance very little, other studies report that member intelligence influences performance a great deal. Therefore, although evidence can usually be found to support a particular thought or concept about leadership, it can usually also be found to support the opposite view. It follows, then, that a very thorough review of the research literature does not lead one to a set of inescapable conclusions or rules regarding those actions a leader ought to take in order to be successful. In other words, whether one is attempting to develop leadership theory which will pass the test of "real-world" use or is just trying to figure out how to do a better job in his present leadership position, he must reject a certain amount of the published research findings. He obviously cannot follow two sets of recommended actions when they tell him to do opposite things.

Academic theorists have also produced a considerable number of leadership "models." The value of a model lies in its ability to accurately tell leaders what they should do in order to achieve good unit performance. For example, if a model says that leaders should take a particular action and we find, when observing "real-world" units, that leaders who take this action are successful and those who fail to take this action are unsuccessful, the model is very useful. But when we find that several leaders who take the recommended action are unsuccessful or that several leaders who fail to take this action are successful, then the model is obviously wrong and of little value. It follows, then, that the best measure of a model's value is its ability to predict group performance. In other words, a model is useful only when there is a strong positive relationship or correlation between its recommended actions and unit performance. When this correlation is low, the model has little value.

A leadership model needs two qualities to predict group performance accurately. First, the model must necessarily include the great majority, preferably all, of the factors which have a significant influence on the outcome. For example, were I to design a leadership model without including leadership style, member ability, and member motivation, three factors which most people feel have strong effects on performance, the value of that model would, most certainly, be low. The problem would be, simply, that my predictions regarding the rise or fall of group performance would not take into account the considerable influence of these three factors. Surely, any predictions based upon a model which fails to include essential components will have a great deal of inaccuracy. Second, each of the model's stated cause and effect relationships, which point out the influence the various included factors have on group performance, must necessarily be correct. In other words, even if my model takes into account all of the essential factors, if my statements or contentions are wrong about the effect each factor has on performance and each of the other factors, the model will obviously be incorrect and have no predictive value. For example, if my model says that leaders will improve unit performance by using a nondirective leadership style whenever they have subordinates of low ability, the model will not be of much use because, as most people agree, the leader who allows incompetent subordinates to perform the task as they see fit is courting disaster. In brief summary, a model which predicts accurately must include most of the factors which significantly influence the outcome and properly describe the cause and effect relationships between the included factors and group performance.

None of the half-dozen leadership models appearing most frequently in current textbooks has achieved wide acceptance because, primarily, they do not predict performance very well. This simply means that when one collects data from "real-world" groups and analyzes it, one finds a considerable number of units in which things do not turn out the way the models predict they will. The principal reason these models do not predict very well is that each of them fails to include a number of factors which both research findings and leadership experience indicate have a significant effect on group performance. For example, the "Contingency Model" does not consider the extent to which the leader is directive or nondirective, the level of the leader's intelligence or ability, or the amount of leader or member motivation. The "Situational Leadership Model" does not include the leader's intelligence or ability, leader motivation, or unit cohesion. Further, neither of these two models accounts for the degree to which the leader enforces performance standards. In summary, none of the published leadership models have achieved much success because each of them has failed to include several factors which significantly affect group performance. This failure lowers the models' predictive ability and, hence, their practical value.

Within the military services, the development of leadership instruction and theory has fared little better. In the precommissioning, junior, and intermediate-level service schools, an observer finds relatively few curriculum hours devoted to the subject, little theoretical substance in the lessons, and a great deal of difference in the content of instruction at the different schools. Lectures, rather than presenting a broad overview of how the important leader and group member factors influence unit performance, too often deal with discussions of what a leader should do in one of the countless number of specific, isolated situations that could exist. It would be far better to teach a set of principles one could apply in any situation than to teach in the present manner. In those schools which do present a certain amount of instruction on leadership theory, the material in almost every case involves a description of one of the academic models discussed earlier.

On another note, when interviewing several commanders at battalion, brigade, and higher levels, one gets very different responses concerning the leadership factors these people consider important. Certainly, most commanders mention some factors such as cohesion and motivation. But, on the whole, they offer many different views on how much emphasis they place on each of the important leadership factors and how a leader should go about improving things such as unit cohesion and motivation. These differences of opinion usually depend upon differences in experience, rather than upon any theoretical basis. Obviously, then, these different commanders give their subordinate leaders a considerable variety of guidance and instruction about how to improve unit performance. Unfortunately, there are many different types of groups and situations in which these subordinate leaders can find themselves and, as this book will show, what works well for someone in one particular situation may work rather poorly in another.

**Purpose**

This book presents an original set of leadership rules to improve unit performance at all organizational levels. The work is designed to fill the considerable gap between the very position-specific, recipe-type lists of "things you should do or not do in this particular job" approach and the "General Patton-type" speech approach that provides a lot of inspiration to excel, but very few specifics as to what one ought to do. In essence, then, this book presents a set of principles or rules that a leader can apply in any situation to improve his group's performance. This work is also intended to serve academic theorists by presenting findings which considerably reduce the great deal of conflict in the research literature and among several of the published theories.

Although I would be uncomfortable with presenting concepts which may prove to be controversial without using the scientific method and including some data as support, it is equally important to me that I do not grossly encumber the field leader, who is interested in practical, common sense ideas, with analyses of variance and academic phrases. Toward that end, I chose to use a rather plain, conversational writing style in order to convey my concepts clearly and to make the book "readable" by the relatively young and inexperienced leader. So as not to ignore the academics' needs, the 10 rules are supported by data gathered from 49 actual Army units and some statistical analyses are presented to support each of my contentions.

## Concepts

Unlike the typical leadership model, which only tells the leader that his style ought to be "task-oriented," "relationship-oriented," "directive," or "nondirective" based upon the presence or absence of two or three particular group factors, this set of 10 rules collectively explains how nondirective leadership, directive leadership, leader enforcement of performance standards, member intelligence, member ability, leader intelligence, leader ability, member motivation, leader motivation, and group cohesion individually influence group performance. Basically, the evidence very clearly establishes that the nondirective and directive leadership styles are each very effective in certain situations and are poor choices to use in others; high levels of enforcement of performance standards, intelligence, ability, motivation, and cohesion will increase unit performance in some situations but will not influence the outcome in others; and low levels of the factors just mentioned will decrease unit performance in certain situations but will not affect the outcome in others. In other words, the effect each of these factors has on performance is not constant but, rather, varies from situation to situation.

The key premise is that the amount of influence each of these factors has on performance depends upon the presence or absence of certain other factors. For example, Rule 5 shows that the influence the group members' ability or job skill has on unit performance depends upon the leader's style and the level of group member motivation. Certainly, if the leader uses a directive style in which he fails to include his subordinates in the job planning and tells them exactly what to do and how to do it, group member ability will have very little effect on the outcome. On the other hand, when the leader is nondirective and pretty much leaves the decisions involving task accomplishment up to the group members, their ability should have a fairly strong influence on performance. Of course, using the nondirective style does not by itself assure good performance. When one's subordinates are talented, allowing them to make their own decisions will result in good performance. However, if one is nondirective with rather incompetent group members, performance will be poor. Similarly, if one's subordinates have low motivation to perform the task, it matters very little if they have high ability or not. Performance will be poor in either case and, thus, unaffected by the members' actual level of ability. But when one's subordinates have high motivation, performance will reflect their talent. This does not mean that high motivation will guarantee a good outcome. Granted, when the members are talented, performance will obviously be good. A great effort by people who don't know what they are doing, however, just won't do very much to improve the end result.

Two key points must be understood from the example in the last paragraph. First, a nondirective leadership style and high member motivation enable group member ability to influence performance. Without the nondirective style or with low member enthusiasm for the task, member ability has little effect on the outcome. Second, the presence of the nondirective style and high member motivation do not by themselves assure good performance. When the members are talented, performance will be good. When they are not talented, performance will be poor. This concept—the influence that leadership style; enforcement of performance standards; member and leader intelligence, ability, and motivation; and group cohesion each have on unit performance depends upon the presence or absence of certain other factors—is used in all 10 of the rules and is the reason why the rules achieved such good results when tested on the sample of 49 Army units.

The format of the book is to present each of the 10 rules as a separate entity. In each case, I develop the rationale, state the rule, present the statistical evidence, and summarize the findings. The next four chapters discuss, in order, the leader actions of style and enforcement of performance standards (Rules 1–3), the group skills of member and leader intelligence and ability (Rules 4–7), group incentive or member and leader motivation (Rules 8 and 9), and group bonding or unit cohesion (Rule 10). In addition to this principal work of the book, separate sections are also included which describe actions that leaders can take in order to raise the level of several important group qualities such as motivation, ability, and cohesion. The summary chapter integrates the separate rules and presents recommended courses of action for the leader to take in order to use the rules properly in whatever particular group situation is encountered. The ability of these rules to improve unit performance in any situation is, of course, their real value.

## Methodology

**A. Subjects.** The data used to test the 10 rules were collected from 49 groups of enlisted men who operated company and battalion-sized Army mess halls at Fort Ord, California. Each group was led by a mess steward (senior noncommissioned officer) and manned by several cooks (lower ranking soldiers). The amount of experience of the mess stewards ranged from 6 to 26 years, while the experience of the cooks ranged from a few months to 24 years. Each of the participants was permanently assigned to work in one particular mess hall while at Fort Ord. Virtually all of the men participated matter of factly, after being assured that their responses were confidential.

The reason I chose Army mess halls as the subjects of this study is that I felt I would be able to obtain a considerably more objective and accurate measure of unit performance from this type of group than from any other type of unit. The most important and difficult task in conducting a field study like this is to obtain an accurate measure of group performance. For most types of military units, it is very hard to measure objectively how good a unit really is at doing its overall job in comparison with similar type units. How does one reliably measure how "good" an infantry platoon, tank company, or basic training battalion "really" is? Typically, when one asks three or four raters, normally, experienced officers assigned at a higher organizational level, to evaluate independently the overall quality of several units, they disagree considerably as to the relative order of merit of the units. Some evaluators rate units based on their opinions of the unit leader's capability, some evaluators rate units based upon how well the unit has scored in the past on several statistical measures, and some evaluators rate based upon how well the unit performed on its most recent field training test.

The problem with conducting a study of units whose performance cannot be accurately measured is obvious. If one does not know how good a unit really is, it is impossible to determine what effect a particular leadership style or a certain level of ability or motivation has on unit performance. This brings us back to the choice of using mess halls as subjects. First, although dining facilities have a different and less "glamorous" mission than do infantry platoons or tank companies, they are certainly still "real-world" groups in every sense of the word, and findings based upon a study of this type of unit are applicable to other types of groups as will be shown later. Second, and more important, I felt that because each mess hall was normally inspected by both brigade and post food service officers on a weekly basis and each inspection was conducted using a written rating form with 12 specific items that were each evaluated on a seven-point scale, the ratings would be considerably more objective than with other types of units, and there would be fairly good agreement among the raters. This proved to be the case. The reliability or agreement between the ratings of the brigade and post food service office evaluators was measured to be $+.76$ or 76 percent, which is, academically, very acceptable for this type of study.

## B. Tests and Questionnaires.

1. *Behavior Descriptions.* In order to determine each mess steward's leadership style, the degree to which he enforced performance standards, and the motivation of both the mess steward and the cooks, each subject was asked to complete a questionnaire. Each question used a seven-point rating scale. The score for each leader or group of members on each of the four behaviors of interest was the average of the individual

scores of the members of that group. The following questions were used to measure the specific behaviors of concern to this study:

a. *Leader Directiveness:* "The mess steward decides what shall be done and how it shall be done."

b. *Leader Enforcement of Performance Standards:* "The mess steward maintains definite standards of performance."

c. *Leader Motivation:* "How hard does the mess steward try to work and do as good a job as possible?"

d. *Member Motivation:* "How hard do you try to work and do as good a job as possible?"

2. *Group Member and Leader Intelligence Scores.* Each subject was asked to complete a 42-item version of the Henman-Nelson Mental Ability Test, a multiple choice instrument containing both verbal and math items. This test has been shown to be a valid measure of intelligence (Buros, 1965) and the 42-item version used in this study had a split-half reliability of + .98 for verbal items and + .76 for math items. Each group's member intelligence score was the average of the individual scores of the cooks within that group, while each group's leader intelligence score was the score of the mess steward of that group.

3. *Group Member and Leader Ability Scores.* Each subject also completed an ability or job skill measure, a 50-item test containing both multiple choice and short answer items. The questions were selected from existing tests administered annually to Army mess hall personnel to measure their job proficiency. The split-half reliability of the test was + .72. Each group's member ability score was the average of the individual test scores of the cooks within that group, while each group's leader ability score was the test score of the mess steward of that group.

4. *Group Cohesion Scores.* Each mess steward and cook also completed a 10-question measure describing the atmosphere of his work group. Typical questions asked whether the group was "friendly-unfriendly," "quarrelsome-harmonious," and "efficient-inefficient." This measure has been used extensively in other research (Fiedler, 1967) to determine the level of leader-member relations. In this study, each group's cohesion score was determined by combining the mess steward's score with the average score of the cooks.

# 2
# Leader Actions: Style and Enforcement of Performance Standards

A thorough review of the research work done in the leadership field indicates rather clearly that if one had to identify the most important of all the factors which have a significant influence on unit performance, the consensus of opinion would be that leadership style is the most critical element. This is not to say that the different researchers and authors agree upon which leadership style should be used in any particular situation, or even upon which leader actions or behaviors should be included in the definition of leadership style. Rather, this simply means these theorists agree that the leader's choice of style is the strongest influencing factor on the amount of success that the group achieves. Examples which support this point include the "Contingency Model," the "Managerial Grid," and the "Situational Leadership Model" or "Life-Cycle Theory of Leadership." In each of these well-known models, the basic premise is that the success of the group depends, in large measure, upon whether or not the leader uses the style that particular model defines as appropriate for his group situation.

## Leadership Style

Without getting into a long discussion concerning the several definitions of style available in the research, I consider leadership style to be the degree a leader tends to be directive or nondirective toward his subordinates. The directive leader keeps most of the decisionmaking and unit control processes to himself; the nondirective, participative leader involves his subordinates in decisionmaking and planning functions and delegates considerable authority to them.

During the last three decades, an increasing emphasis has been placed on the participative leadership approach, both in theory and everyday practice. In particular, *In Search of Excellence* (1982), *Management By Objectives* (1954), and work on Situational Leadership in one way or another each stress the desirability of using the nondirective leadership style, at least in certain group situations. A number of research studies support this notion. For example, Coch and French's (1948) well-known study of factory workers found that subordinates who had a say in how things were to be done produced far better results than did workers who had no say. In all of these writings, the authors advocate using the nondirective leadership style because it takes advantage of group member knowledge, tends to motivate the group members, and ensures that members clearly understand how to do the task.

But other studies and examples suggest that the directive style is either just as effective as the participative approach or is, in some cases, even more effective. For example, Stogdill, in his *Handbook of Leadership* (1974) writes, "group productivity does not vary consistently with directive or participative styles." In support of the directive style side of the argument, it certainly makes sense that one would not want to be a nondirective, participative leader if his subordinates were relatively unskilled or did not know very much about the task at hand. The drill sergeant does not ask the advice of soldiers in basic training about how they should be trained nor does the pilot seek the opinions of his crew as to how best he should fly the plane. In both cases, the group members simply do not have the knowledge necessary to make suggestions of much value.

Thus, on the one hand, evidence indicates that participative leadership is the best approach, while on the other, the directive style seems to be the most effective. I believe the key needed to resolve this argument and determine what influence leadership style actually has on performance lies in my fundamental concept that the influence a principal factor has on performance depends upon the presence or absence of certain other factors. In the case of leadership style, the evidence suggests that member ability, member motivation, and leader ability are the relevant other factors. Ability, of course, refers to a person's job skills and knowledge of the task at hand.

At this point, I need to explain how the discussion of the effect of leadership style on group performance will be organized. The relationship between style and performance is somewhat more complex than are the other relationships addressed in this book and must be viewed from the separate perspectives of member ability and leader ability. Under the nondirective leadership style, which relies considerably on subordinates' skills, member ability should certainly have a stronger effect on performance than it would under the directive style, which prevents much member participation. In other words, when the members have a major role in deciding how the task will be done, performance depends much more on their abilities than if they had no say. Under the directive style, which relies considerably on leader skills, leader ability should have a stronger effect on performance than it would under the nondirective style, in which the leader contributes little to the outcome. When the leader uses his own expertise to decide how to do the task, performance depends much more on his abilities than if he lets the members decide how to do the task on their own. Figure 1 illustrates these relationships.

In Figure 1 the horizontal axis measures the effect ability has on performance, while the vertical axis measures the leaders's style. The two diagonal axes represent member and leader ability. Thus, when the leader's style is very nondirective (A), member ability has a strong effect on performance and leader ability has little effect on the outcome. But when the leader's style is very directive (B), member ability has little effect on performance and leader ability has a strong effect on the outcome. Further, when the leader's style is about equally nondirective and directive (C),

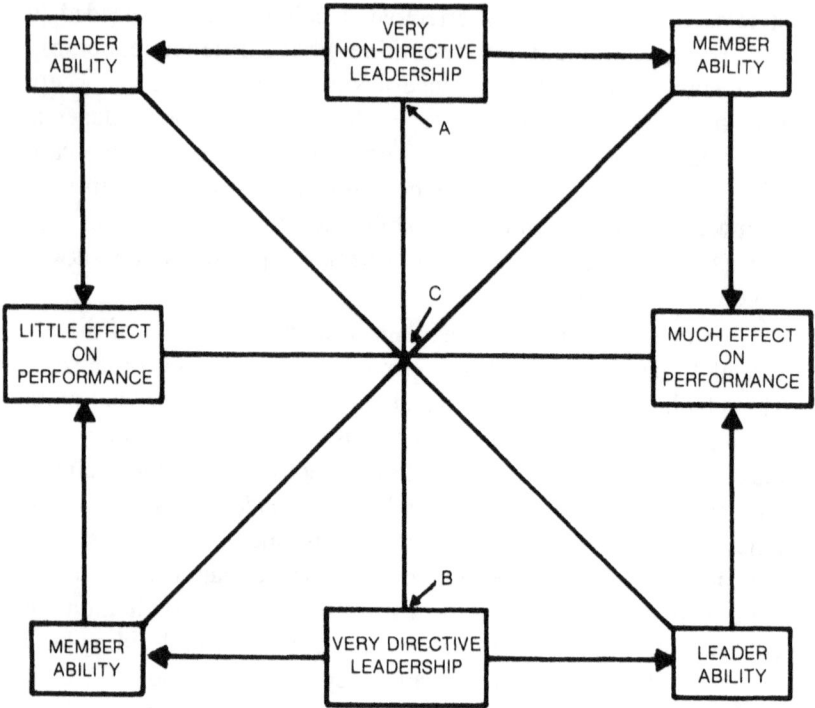

*Figure 1. How Differences in Leadership Style Affect the Influence of Member and Leader Ability on Performance*

member and leader ability each have a medium amount of influence on performance. The next few pages clearly show that the nondirective and directive leadership styles are not competing to see which one is "correct." Rather, the supporting evidence indicates that each style is effective in different situations and the two styles work in a complementary, rather than conflicting, manner. The section summary explains the combined effect.

Perhaps the clearest way to discuss the effect of leadership style on performance is to explain it within the context of an actual example. While this example is that of a battalion commander leading subordinate companies, the principles are the same regardless of the organizational level. The basic training battalion to which I was assigned had four companies: Alpha, Bravo, Charlie, and Delta. Because this example describes a setting and events which actually happened and which could, therefore, be potentially embarrassing to real people, I have changed both the names and the relative order of the companies. Thus, with the names that I use, Echo, Foxtrot, Golf, and Hotel, Echo is a company other than the real Alpha. During my first 30 days in the unit, I spent a great deal of time observing day-to-day training and the abilities and motivation of the unit officers and noncommissioned officers. While I had not previously been assigned to a unit with this particular mission, I felt I knew, reasonably well, how to train initial entry soldiers because prior to my assumption of command, I had spent a great deal of time talking to incumbent commanders, studying, observing training, and attending the Pre-Command Course for new battalion commanders. After one month, my evaluation of my subordinate leaders was (a) Foxtrot and Hotel both were high in ability and high in motivation, (b) Echo was high in ability but low in motivation, and (c) Golf was high in motivation but low in ability. With this as the setting, let us discuss the effect of leadership style on performance from the perspective of member ability, remembering that in this example I am the leader and the company commanders and their drill sergeants are the members of my group.

**From the Perspective of Member Ability.** Member ability should have a fairly strong effect on group performance under nondirective leadership because this style relies considerably on the members' talents. On the other hand, under directive leadership, member ability should have little effect on the outcome because this style relies on leader skills rather than on those of the members. It follows that if the company commanders and their drill sergeants (the cadre) of a basic training battalion have both the technical knowledge and teaching skills necessary to train the new soldiers well in rifle marksmanship, physical readiness, drill and ceremonies, and the other required skills, nondirective leadership would be a good choice to use. This style assures that the cadre's high ability plays a major role in soldier training and, hence, causes good performance. But when the cadre does not have the skills necessary to do a good job training the new soldiers, it obviously would be a mistake for the leader to use the nondirective style. In this case, the limited talents of the cadre would have a strong influence on soldier training and unit performance would be poor. Thus, high member ability will cause the nondirective style to be effective and low member ability will cause the nondirective style to be ineffective. Further, the effectiveness of the directive style is not related to the level of member ability because this style prevents the members' talents from having much influence on the outcome.

Member motivation to accomplish the task should also affect the amount of influence leadership style has on performance. Given the company commanders and drill sergeants are competent to train the new soldiers and the battalion commander uses the nondirective style, surely the amount of effort the cadre is willing to put forth will affect the outcome. With talented cadre members who try hard, performance should be good. However, if the cadre merely "goes through the motions" of working and does not expend enough effort to do the job right, how can the nondirective style be effective? It is no more useful in terms of performance to have talented subordinates who will not make a decent effort than it is to have untalented members. With either group, performance is going to be poor.

I am suggesting, first, that when the leader has competent and motivated subordinates, the nondirective, participative style will produce good unit performance. This style takes advantage of high member ability and motivation by including the members in planning and decisionmaking functions and by delegating considerable authority to them. Thus, in the basic training battalion, it would appear that nondirective leadership is a good choice to use with Foxtrot and Hotel. The considerable talents and enthusiasm of these companies will assure that the new soldiers of these two units are trained well. Second, when one's subordinates have little talent or are poorly motivated, the nondirective leadership style will produce poor unit performance. It just doesn't make good sense to allow the members to have a strong influence on group performance when they don't know what they are doing or don't care enough to do a good job. At this point, then, it appears that I certainly do not want to be nondirective with Golf, which has low ability, or Echo, which has low motivation, because this style will guarantee that the new soldiers of these two companies receive poor training. Finally, the effectiveness of the directive leadership style has little to do with member ability because this style relies on the leader's talents and prevents member abilities from having much effect on performance. Therefore, the directive style may prove to be a good alternative for leading Golf and Echo.

*Rule 1: When the members have high ability and high motivation to accomplish the task, the nondirective leadership style will produce above-average group performance. When the members have low ability or low motivation, the nondirective leadership style will produce below-average group performance. The effectiveness of the directive leadership style is not related to member ability.*

It is important to understand how to read the tables because they offer clear evidence that the 10 rules presented in this book did, in fact, hold true when tested on 49 Army units. Although I would have preferred to present correlational evidence in all 10 tables, I was only able to do so for the last 8 tables. For Tables 1 and 2, a comparison of average group performance scores was the best available method of analysis. In studies such as this, one really ought to have a minimum of 9 or 10 groups in each category (for example, the category of high member ability/high member motivation) in order to calculate meaningful correlations. Achieving this number of groups was no problem for the last eight tables. However, for Tables 1 and 2, the sample of 49 Army units had to be categorized or divided by three factors instead of only the two factors needed for Tables 3–10. For example, in Table 1, I had to group units according to the leader's style, member ability, and member motivation. As expected, this resulted in having only six groups in the nondirective-led/high member ability/high member motivation category. Thus, for the first two tables, meaningful correlations could not be calculated and the best analysis method was to compare average group performance scores.

**Table 1.** Comparison Between the Performance Scores of Nondirective-Led Groups

| Category | |
|---|---|
| *1* | *2* |
| High Member Ability and High Member Motivation | Low Member Ability and/or Low Member Motivation |
| X̄ = +1.213 | X̄ = −.162 |
| % = top 29% of all 24 groups | % = bottom 42% of all 24 groups |

X̄ = Average performance score for groups in category

% = Position of X̄ in range of all 24 nondirective-led group performance scores

The test of Rule 1 was conducted in the following manner. First, the entire sample of 49 Army units was categorized based upon the levels of leader directiveness, member ability, and member motivation. The score used to categorize the groups as high or low on each factor was the middle score for all 49 groups. In other words, groups scoring above the average on a particular factor were rated as high on that factor, while groups scoring below the average were rated as low. Groups with the exact middle score were not included. Second, using only the groups with nondirective leaders, I distributed each group to one of the two categories of Table 1, based upon its levels of member ability and motivation. In each of these two categories, I then calculated the average performance score, represented by the symbol $\bar{x}$, for the groups which fell into that category. I then determined where that average score fell on the range of all 24 nondirective-led group performance scores. This is represented by the symbol %.

The findings in Category 1 of Table 1 show that, as predicted, nondirective leadership produces above-average group performance when the members have high ability and high motivation. The average performance score of the 6 groups in this category was in the top 29 percent of all 24 nondirective-led groups. Also, as shown in Category 2, with members of low ability or poor motivation, nondirective leadership produces below-average group performance. The average performance score for the 18 groups with these types of members was in the bottom 42 percent of all 24 nondirective-led groups. Clearly, there is little doubt that the nondirective style was much more effective with competent, enthusiastic members than it was with less-talented, poorly motivated subordinates.

*In summary of Rule 1, there are three key points to remember. First, if one has talented and enthusiastic group members, the nondirective leadership style will produce good group performance. This style is effective because the nondirective leader takes advantage of his subordinates' considerable knowledge and enthusiasm by allowing them a major role in the planning and decisionmaking functions and by delegating authority to them. Sec-*

*ond, if the members have little ability or poor motivation, nondirective leadership will produce poor group performance. Allowing subordinates to have a major role in the task when they have limited talents or aren't willing to put forth much effort is, obviously, unwise. Third, the effectiveness of the directive leadership style is not related to the level of member ability.*

**From the Perspective of Leader Ability.** Clearly, leader ability is capable of influencing unit performance. It is a widely accepted practice in both military and civilian organizations to select the most competent people for responsible leadership positions in the belief that a group led by a talented leader will achieve more than will a group led by a less-talented leader. The leader's skills influence performance in a different manner than does his subordinates' ability. In the basic training battalion, while the cadre members' ability causes unit performance to be better or worse depending upon how well the cadre trains the new soldiers, the battalion commander's ability affects performance through his role as the adviser and trainer of the cadre members. It is the commander's job, by using his knowledge and experience, to give his cadre members better ideas, methods, and techniques to use in training new soldiers. Of course, the leader may or may not feel confident enough to do this. It follows, then, that when the leader is nondirective and fails to contribute his guidance to his subordinates, unit performance will not be influenced by the leader's ability. However, when the leader is directive and gives his advice to the cadre, performance will be considerably influenced by the quality of his suggestions. The talented leader, whose good ideas improve his subordinates' skills, will cause good performance. The incompetent leader, who requires his cadre to follow his rather inept guidance, will cause poor performance. Thus, high leader ability will cause the directive style to be effective and low leader ability will cause the directive style to be ineffective. Further, the effectiveness of the nondirective style is not related to the level of leader ability because this style prevents the leader's talents from having much influence on the outcome.

As with nondirective leadership, the amount of influence the directive style has on group performance must be affected by the level of member motivation. If a talented battalion commander directively provides excellent advice to his subordinates, surely the amount of effort the cadre members are willing to make in using this guidance will affect performance. Subordinates who work hard and follow good advice will perform well. However, if the cadre does not enthusiastically carry out the commander's guidance and, instead, chooses to do just enough work to stay out of trouble, how can the directive style be effective? It is no more useful in terms of performance to provide good suggestions to unmotivated subordinates than it is to provide poor suggestions. In either case, performance is going to be low. The fact that the directive leader supervises his subordinates more closely than does the nondirective leader does not ensure a more productive effort. A leader cannot supervise in more than one place at a time nor does admonishing subordinates always serve to elicit more than the reaction of "looking a little busier."

I am suggesting, first, that when the leader is talented and has motivated subordinates, the directive leadership style will produce good unit performance. This style takes advantage of the leader's considerable skills and the members' enthusiasm to carry out his instructions. Thus, in the basic training battalion, it would appear that directive leadership is a good choice to use with Golf. The combination of my adequate guidance and Golf's enthusiasm to put this advice to good use will assure that the new soldiers of that company are well trained. Second, when the leader has little ability or his subordinates are poorly motivated, the directive leadership style will produce poor unit performance. Being directive when one doesn't know what he is talking about is not very smart. Further, being directive when one's subordinates are not willing to carry out their leader's instructions is of little value. Thus, it seems I have a problem on my hands with Echo. With that company's low motivation, it appears that neither the directive nor the nondirective style will be effective. Third, the effectiveness of the nondirective leadership style has little to do with leader ability because this style relies on the members' talents and prevents leader abilities from having much influence on performance.

*Rule 2: When the leader has high ability and the members have high motivation to accomplish the task, the directive leadership style will produce above-average group performance. When the leader has low ability or the members have low motivation, the directive leadership style will produce below-average group performance. The effectiveness of the nondirective leadership style is not related to leader ability.*

The same method used earlier to test Rule 1 was used here to test Rule 2. The findings in Category 1 of Table 2 show that, as predicted, directive leadership produces above-average group performance when the leader has high ability and the members have high motivation. The average performance score of the groups in this category was in the top 8 percent of all 24 directive-led groups. Also, as shown in Category 2, with leaders of low ability or members with poor motivation, directive leadership produces below-average group performance. The average performance score for the groups with these types of people was in the bottom 42 percent of all 24 directive-led groups.

*In summary of Rule 2, there are three key points to remember. First, if the leader is talented and has enthusiastic subordinates, the directive leadership style will produce good group performance. The directive approach is effective because it enables the bright leader to contribute his good ideas and knowledgeable techniques to the work effort and, hence, improve performance. Second, if the leader has low ability or the members have poor motivation, directive leadership will produce poor group performance. It just isn't a good idea for the leader to have the major influence on performance when he has limited skills. Also, when his subordinates are not willing to wholeheartedly carry out the leader's guidance, being directive does nothing to improve the outcome. Third, the effectiveness of the nondirective style is not related to the level of leader ability.*

Summarizing both Rules 1 and 2, the findings very clearly indicate which style the leader ought to use for any of the possible situations he could find himself in when the situation is defined by

**Table 2.** Comparison Between the Performance Scores of Directive-Led Groups

| | Category | |
|---|---|---|
| | *1* | *2* |
| | High Leader Ability and High Member Motivation | Low Leader Ability and/or Low Member Motivation |
| | $\bar{x}$ = +2.139 | $\bar{x}$ = −.688 |
| | % = top 8% of all 24 groups | % = bottom 42% of all 24 groups |

$\bar{x}$ = Average performance score for groups in category

% = Position of $\bar{x}$ in range of all 24 directive-led group performance scores the levels of member and leader ability. Given, of course, that the members are motivated, if the leader is talented and the members are not, the correct choice is the directive style. This style ensures that the leader's bright ideas and knowledgeable guidance play a major role in unit performance. In this case, the members' lack of ability will not be very detrimental because their role in making decisions is minimal. However, if the leader was nondirective in this situation, his skills would not be utilized and the members would have to rely on their own poor abilities. Obviously, performance would suffer. In the basic training battalion example, at the end of my first 30 days of command, this was the exact situation that I hoped I was in with Golf Company. I knew Golf was very enthusiastic but not very talented. I was hoping I had learned enough to be competent because Golf obviously needed me to be directive and provide them with good advice and guidance concerning methods and techniques for training new soldiers. To be nondirective and leave them to fend for themselves clearly would not work. Fortunately, I had apparently learned enough because the advice I gave to Golf improved their performance.

On the other hand, if the members are talented and the leader is either not very bright or hasn't yet learned very much about his new job, the nondirective style is the right choice. This style puts high member abilities to good use because the members play a major planning and decisionmaking role. At the same time, the leader's lack of ability or experience has little effect because his is not the dominant voice. Were this leader to be directive, performance would reflect his poor talent or inexperience rather than the excellent skills of the members. In the basic training battalion, I was not really sure, at the end of my first 30 days, whether or not this was the situation I was in with Foxtrot and Hotel. I thought that they were bright and motivated, but I just wasn't sure whether I knew as much as I hoped I did. I felt that I couldn't lose if I was nondirective with these two companies because under this style, performance would reflect their excellent talents. The directive style offered no advantage and considerably more risk because of my uncertainty regarding my own skills. The choice of the nondirective style worked out fine.

In the case where neither the leader nor the members have very much ability, there's "big trouble in River City." Regardless of whether the leader is directive and relies on his own poor skills or whether he is nondirective and relies on the poor skills of his subordinates, performance is going to be poor. This leader needs to learn the required skills in a hurry. In the basic training battalion, I was hoping that I was not in this position with Golf. I knew that Golf had poor skills and I hoped that I didn't. Fortunately, this was not the situation because I apparently had learned enough to pass on some useful advice to them. However, this example points out to incoming leaders the great value of preparing one's self well for command.

The choice of which leadership style to use with groups composed of leaders and members who both have high ability is a matter of preference rather than necessity. In this enviable situation, either the nondirective or the directive style will produce good unit performance. The first will take advantage of high member skills, and the second will use the leader's good talents. It is worth noting that when the leader is bright, his competent and enthusiastic subordinates apparently do not mind a directive style. This is understandable because, in this situation, the directive approach produces good unit performance and that, of course, is the primary goal. However, while either style will be effective, the nondirective style has the added benefit of serving to increase member motivation and is, therefore, the more preferable choice. In the basic training battalion, this was the situation I hoped I was actually in with Foxtrot and Hotel. I knew that they were competent and I hoped that I was. While, from hindsight, either leadership style would have been effective, I chose the more certain route of nondirective leadership because performance would reflect the considerable skills which these companies had demonstrated to me.

The recommendations that have just been made about the choice of a particular leadership style in any specific leader-member ability situation assume the group members are highly motivated to accomplish the task. When that motivation is not present, neither leadership style has a strong, positive influence on performance and, thus, there is no valid basis upon which to recommend a particular leadership style in a specific setting. The dangers of trying to lead a unit while having no idea of which leadership style will be most effective are readily apparent. Unfortunately, this was the situation I found myself in with Echo Company. Echo was talented, but just didn't put very much effort into its work. Regarding ability alone, either leadership style would have been effective because they were competent as was I. But, until they became motivated, neither style would improve their performance. Obviously, then, my first challenge with this company was to improve the cadre's motivation. Chapter 4 discusses, in some depth, how the leader can improve member motivation, but the short answer here is that the combination of using the nondirective style, which tends to increase talented members' motivation, the challenge of winning the newly instituted "Best Company of the Cycle" award, and the considerable personal enthusiasm I displayed for the job of training the new soldiers eventually got Echo "fired up" and performing well.

### Enforcement of Performance Standards

This section addresses the effect the leader action of enforcing performance standards has on group productivity. The leader is defined here as enforcing high standards when he sets a high minimum level of acceptable performance for his group members and is very insistent that his subordinates meet that acceptable level. It is important to distinguish between the concepts of enforcing standards and leadership style because these two leader actions are often confused. On the one hand, leadership style focuses, primarily, on who is involved in the planning and decisionmaking of the task. On the other hand, enforcement of standards emphasizes the degree to which the leader requires high quality work from his subordinates. Thus, the level to which the leader enforces standards does not depend upon which leadership style he uses or vice versa. In the basic training battalion example, while I used different leadership styles based upon each unit's capability, I enforced the same performance standards in all four companies.

The military service has long recognized "the enforcement of standards" as a vitally important leadership principle. An example paraphrased from Sun Tzu's *The Art of War* clearly illustrates the positive effect which the enforcement of high standards can have on performance. Sun Tzu, by means of his book on the art of war, secured an audience with Ho-lu, the King of Wu. Ho-lu said, "I have read your 13 chapters, Sir. Can you conduct a minor experiment in control of the movement of troops?" Sun Tzu replied, "I can." Ho-lu asked, "Can you conduct this test using women?" Sun Tzu said, "Yes." The King thereupon agreed and sent from the palace one hundred and eighty beautiful women. Sun Tzu divided them into two companies and put the King's two favorite concubines in command. He then said, "Do you know where the heart is, and where the right and left hands and the back are?" The women said, "We know." Sun Tzu said, "When I give the order 'Front,' face in the direction of the heart; when I say 'Left,' face toward the left hand; when I say 'Right,' toward the right; when I say 'Rear,' face in the direction of your backs." The women said, "We understand." After these regulations had been announced, the executioner's weapons were arranged to make it clear Sun Tzu meant business.

Sun Tzu then gave the orders three times and explained them five times, after which he beat on the drum the signal "Face Right." The women all roared with laughter. Sun Tzu said, "If regulations are not clear and orders not thoroughly explained, it is the commander's fault." He then repeated the orders three times and explained them five times, and gave the drum signal to face to the left. The women again burst into laughter. Sun Tzu said, "If instructions are not clear and commands not explicit, it is the commander's fault. But when they have been made clear, and are not carried out in accordance with military law, it is a crime on the part of the officers." Then he ordered the commanders of the right and left ranks beheaded as an example. He then used the next senior as rank commanders. Thereupon, he repeated the signals on the drum and the women faced left, right, to the front, to the rear, knelt, and rose all in strict accordance with the prescribed drill. They dared not to make the slightest noise. Ho-lu, although very upset by the loss of his favorite concubines, realized Sun Tzu's capacity as a commander and eventually made him a general. Sun Tzu won many victories and raised the name of Wu to an illustrious position among the feudal lords of China.

Although I am not advocating summary execution for mishaps on the drill field, I believe leaders who establish high levels of acceptable performance and who refuse to accept anything less will get exceptional results from their subordinates. The reason the high enforcement approach is more effective than the low enforcement approach is simply that the first is able to put into use more of the group members' skills and motivation. It may take several practice runs before one's subordinates can do the task well enough or before they understand that the leader will take nothing less, but eventually they will do it right. As one might expect, the leader who sets lower performance levels and who routinely accepts less from his subordinates will normally get less.

As with leadership style, it is reasonable to believe both member ability and member motivation will affect the influence that enforcement of standards has on group performance. Unless the members have sufficient skills and enthusiasm to accomplish the work, the leader's attempts to enforce high standards will most likely be futile. In the case of ability, Sun Tzu clearly knew that unless he gave explicit instructions and trained his subordinates to carry them out, he should not expect performance to reach his high standards. As a modern day example, drill sergeants do not get upset with new soldiers who, during their first day of basic training, cannot execute a perfect "Right Face." The sergeants understand that, without practice, the trainees cannot meet the required standard. The leader who berates subordinates who do not know how to carry out his orders would more profitably spend his time training them. With low member ability, then, the degree to which the leader enforces standards will have little effect on performance.

In the case of motivation, Sun Tzu understood that unless his subordinates were willing to try their best to accomplish the task, it mattered little whether he attempted to enforce high standards or not. He could keep the women out on the drill field all night long, continuously repeating the orders to them and refusing to accept their laughter in place of the correct drill movements, but the women's performance would not improve until they were willing to make the necessary effort. In basic training, every new soldier, in order to graduate, must meet the established high standards. For those trainees who try hard, the drill sergeant will make every extra effort to help them meet the standards. For trainees who are not willing to make a good effort, the drill sergeant has little time. He knows that the standards will not be lowered and, despite his best efforts, trainees who do not try hard will not reach the acceptable level of performance. Thus, with low member motivation, the degree to which the leader enforces standards will have little effect on performance.

I am suggesting that leader enforcement of performance standards will have little effect on group productivity when the members have little talent or are poorly motivated. Thus, in my basic training battalion example, I initially had a problem with Golf, which had low ability, and Echo, which had low motivation. On the other hand, with capable and enthusiastic members, enforcement of high standards will have a positive influence on performance, while failure to enforce high standards will have a negative influence on the outcome. Specifically, when the leader enforces high standards, he causes the members to use all of their considerable ability and motivation in order to reach the required performance level. When the leader sets relatively low standards, however, the members need only use a portion of their capabilities and enthusiasm to reach those standards and, normally, the performance a leader gets is only as good as what he requires. With the talented and motivated members of Foxtrot and Hotel, then, one would expect that enforcing high standards would favorably influence performance.

An example which illustrates these points occurred during my fourth week of command when the battalion as a whole practiced for the Command Retreat parade ceremony. I told the cadre at the start of practice that this particular ceremony was important for several reasons, and we would practice until the battalion could perform the ceremony exceptionally well. After two rehearsals,

Foxtrot and Hotel looked excellent, but Golf and Echo were poor. I dismissed Foxtrot and Hotel and put Golf and Echo through two more rehearsals. Much to my dismay, the results were only marginally better. The problem with Golf, which had low ability, was clearly that it needed more practice. I gave Golf that opportunity. The problem with Echo was that its cadre members just didn't care very much whether they looked good or not. Because the ceremony was only two days away and I felt that methods of improving Echo's motivation on a permanent basis would take considerably longer than that, Echo's cadre and I had a rather harsh, one-sided "conversation" that evening concerning the responsibilities of the cadre members. As I hoped, that approach worked for the necessary two days.

*Rule 3: When the members have high ability and high motivation to accomplish the task, the more the leader enforces high performance standards, the better group performance will be. When the members have low ability or low motivation, leader enforcement of high standards will have little effect on group performance.*

The tests of Rules 3–10 are in the form of correlations. I was able to use this preferable test method because, for these rules, the sample of 49 Army units needed only to be categorized by two factors as opposed to the three factors needed to test Rules 1 and 2. This enabled each category (for example, high member ability/high member motivation) to contain the 10–12 groups needed to calculate meaningful correlations.

Let me explain how the test of Rule 3 was conducted. For consistency, this same test is the one used for every remaining rule in this book. First, the entire sample of 49 groups was divided into four categories, based upon the levels of member ability and member motivation. As done before with Rules 1 and 2, the score used to categorize the groups as high or low on each factor was the middle score for all 49 groups. Again, groups with the exact middle score were not included. Second, the groups were distributed to Table 3. The symbol *n* indicates the number of groups which fell into each category. Third, correlations were computed between group performance and the degree to which the leader enforced

**Table 3.** Correlations Between Leader Enforcement of Performance Standards and Group Performance

| | Category | | |
|---|---|---|---|
| *1* | *2* | *3* | *4* |
| High Member Ability High Member Motivation | High Member Ability Low Member Motivation | Low Member Ability High Member Motivation | Low Member Ability Low Member Motivation |
| n = 12 groups r = +.545* | n = 9 groups r = −.217 | n = 13 groups r = +.115 | n = 10 groups r = +.342 |

n = Number of groups in category

r = Correlation between the degree to which the leader enforced high standards and group performance

* = Statistically significant at .05 level (one-tail)

high performance standards. These correlations are indicated by the symbol $r$.

Although I am trying, for the convenience of the reader, to avoid statistical details as much as possible, there are a few points concerning correlations that are important for the reader to understand. Understanding the "numbers" in the tables is important because they are solid evidence that the rules I present really did hold true when tested by the data collected from 49 actual Army units.

First, a positive sign in front of the correlation means that the more the leader enforces high standards, the better group performance tends to be, and the less the leader enforces high standards, the worse performance tends to be. A negative sign in front of the correlation means just the opposite: the more the leader enforces high standards, the worse performance tends to be, and the less the leader enforces high standards, the better performance tends to be.

Second, the three numbers following the positive or negative sign indicate the strength of the correlation. The numbers can range from a low of .000 to a high of 1.000. The higher the size of the number, the stronger the relationship between leader enforcement of high standards and group performance. In other words, the correlation of .545 in Category 1 of Table 3 indicates that leader enforcement of high standards has a fairly strong effect on performance in this situation. In contrast, the correlation of .115 in Category 3 of Table 3 indicates that leader enforcement of high standards has almost no effect on group performance in this situation.

Third, the presence of one asterisk beside a correlation indicates that the correlation is statistically significant. This simply means that I could not have gotten a correlation this high by luck or chance more than once in 20 times. Academically, this rate is very acceptable and, of course, adds considerable credibility to the finding. A double, triple, or quadruple asterisk means that a chance correlation this high would only occur once in 40 times, once in 100 times, or once in 1,000 times, respectively.

Fourth, the tables in this book contain several nonsignificant correlations in the ±.200–.500 range. An example, +.342, is in Category 4 of Table 3. It needs to be understood that these correlations, because they are not statistically significant, might well have happened by chance and that even if these correlations were not a matter of luck, the strength of the relationship between the factors involved is rather weak and, therefore, of little consequence.

Finally, I must point out that a correlation does not mathematically prove causality. In other words, the fact that two items correlate is not proof, by itself, that the movement of the one item is definitely the cause of the corresponding movement of the other. Certainly, it is a prerequisite for two items to correlate in order for one to influence the other. But, the rational "proof" that one item does, in fact, directly influence the other is established, as I will do throughout the book, by additional evidence in the form of research studies, relevant everyday examples, or good common sense.

The findings in Category 1 of Table 3 show that, as predicted, leader enforcement of high performance standards has a positive influence on group productivity when the members are talented and motivated. Also, as shown in Categories 2, 3, and 4, for groups with members of little ability or motivation, leader enforcement of high standards has little effect on performance. Perhaps the practical-use value of the correlations in Table 3 is more clearly illustrated for some by a comparison of the average group performance scores, as was done in Tables 1 and 2. Of the 12 groups in Category 1 of Table 3, seven of the groups were led by leaders who enforced high standards and five by leaders who did not enforce high standards. The groups with leaders who enforced high standards had an average performance score in the top 39 percent of all 49 groups, while the groups with leaders who did not enforce high standards had an average performance score in the bottom 10 percent of all groups. The average performance scores of the groups in Categories 2, 3, and 4, which had members of low ability or poor motivation, ranged between a low of the 35th percentile to a high of the 58th percentile.

*In summary of this section, there are two key points to re-*

*member. First, if one has capable and enthusiastic subordinates, the more the leader enforces high performance standards, the better his unit will tend to perform. The reason the high enforcement approach is more effective than the low enforcement approach is simply that the first uses a greater amount of the group members' skills and motivation. As a general rule, some subordinates always try their best, while some choose only to do an amount "that's good enough for government work," and still others do just enough to stay one step in front of "the law." The leader who enforces high standards will get a lot more work from these latter two groups than will the leader who accepts whatever he gets. Second, if one has subordinates of little ability or motivation, the degree to which the leader enforces standards has very little influence on performance. Because the findings show that leader enforcement of high standards is capable of paying definite performance dividends, it is clearly worthwhile for the leader to devote considerable attention to the training and motivation of his group members.*

# 3
# Group Skills: Member and Leader Intelligence and Ability

The issue of how much influence, if any, that member and leader intelligence and ability have on unit performance is perhaps the most controversial of any of the relationships discussed in this book. Not only do the numerous research studies on this subject indicate considerable interest but widely disparate findings have fueled the debate. Equally important, the effects of intelligence and ability on performance have great implications in the everyday, "real" world. For example, those responsible for making the hard decisions regarding national defense are keenly concerned with this issue. Defense planners must decide whether the considerable costs of pay, manpower, and time devoted during the last five years to recruit talented people into the military are worth that effort or whether a military composed of the same number of people although with lower mental entrance scores is just as effective at considerably less cost. The answer, of course, lies in how much influence talent has on performance.

## Member Intelligence

I distinguish between intelligence and ability by considering the first to be the talent one uses primarily during the planning and decisionmaking phase of the task. A person uses his intelligence to figure out what alternatives are available to him and to decide which option is the best choice. On the other hand, ability refers to the specific skills and task-relevant knowledge required of people during the performance phase to actually accomplish the work. In order for the new soldier to hit a target with a rifle, the drill sergeant must have the marksmanship knowledge and teaching skills to train the soldier, and the soldier must acquire the skills necessary to shoot the rifle correctly. This is not to say that a "hard and fast" line exists so that a person uses absolutely no intelligence during task performance or no ability during task planning. Certainly, there is some amount of overlap. But, on the whole, the organization and planning of a task typically uses a different set of talents than those used for actually performing the work. Therefore, it follows that intelligence and ability are not competing to determine which one will influence performance. Rather, both can contribute at different times during the task.

Support for the view that high member intelligence has a positive influence on performance can be found in discussions of two manpower accession decisions. As conscription gave way to voluntary service in the early 1970s, the Army found that although increased pay and allowances were generally able to fill the ranks with sufficient numbers of soldiers, the percentage of soldiers from the upper mental categories was dwindling. The problem was aggravated by the fact that the mental entrance examination was inadvertently mis-normed so that a large number of soldiers who would not otherwise have been eligible for service were allowed to enlist. In 1979, the Army and Congress recognized that soldier intelligence has a considerable positive effect on performance. Faced on the one hand with an influx of new equipment and systems requiring talented people and on the other hand with large numbers of marginally capable soldiers, the Army raised its mental entrance requirements. At the same time, Congress reduced the number of lower mental category soldiers that could enlist. The rationale behind these actions was clearly that bright people learn faster, learn more, and retain more, and these qualities result in better performance.

A second example, Project One Hundred Thousand, also supports this point. From 1966 to 1969, the military services lowered entrance requirements in order to accept some 246,000 servicemen who would not otherwise have been qualified for enlistment. Of these recruits, 92 percent were previously unqualified by reason of low intelligence scores. A 1969 Office of the Secretary of Defense report on this project summarized, "As could be expected, the men brought in under reduced mental standards do not perform as well as a cross section of men with higher test scores and educational abilities. This is true on all measures—training attrition, promotions, supervisory ratings, disciplinary records, and attrition from service."

It is reasonable to assume that leadership style and member motivation will affect the influence member intelligence has on performance. Regarding leadership style, the primary issue is how much opportunity member intelligence has to contribute to the task. Surely, when the leader's style is directive and the members have little participation in the planning and decisionmaking functions, member intelligence should affect performance very little. On the other hand, the success of the nondirective leader, who allows the members to participate in the planning discussions and who relies on the advice of his subordinates, will depend to a considerable extent on the quality of their input. In the basic training battalion example, my nondirective style with Echo, Foxtrot, and Hotel allowed these three companies the chance to have a considerable say in how they performed a lot of their training. The excellent talents of these units had a strong, positive influence on performance and would have been wasted had I been directive and not asked for their opinions. But with Golf, which was not very capable, I chose to be directive because it just doesn't make good sense to take advice from people who have little talent. Thus, when the leader's style is directive, member intelligence will have little effect on performance. On the other hand, when the leader is nondirective, member intelligence will have a fairly strong influence on the outcome.

Regarding member motivation, it seems obvious that highly intelligent people who are not willing to expend very much effort will not improve an organization's performance. What is the point of inviting a competent company commander to a planning meeting if he just sits there and says nothing? This was the problem I had initially with Echo's commander. Although the Foxtrot and Hotel commanders readily offered good suggestions for better methods to train the new soldiers, Echo's commander was content to sit and say virtually nothing. He responded to direct questions, most frequently, with "Yes," "No," or "I agree" answers. Thus, his good talent, unfortunately, contributed little to the task. Therefore, when one's subordinates are poorly motivated, their intelligence will have little effect on performance. When subordinates are enthusiastic, however, their intelligence will have a fairly strong influence on the outcome.

I am suggesting that member intelligence will have little effect on group performance when the leader uses a directive style or the members are poorly motivated. Thus, because I was directive with Golf, that company's intelligence had little influence on the outcome. This style was a good choice, though, because Golf had little talent. Echo's intelligence also had little early effect on performance because that company's cadre was not motivated. This was unfortunate because Echo was competent and could have made a good contribution. I am also suggesting that when the leader's style is nondirective and the members are highly motivated, high member intelligence will have a positive influence on performance and low member intelligence will have a negative influence on the outcome. Thus, with Foxtrot and Hotel, one would expect their considerable talent to produce excellent performance.

*Rule 4: When the leader's style is nondirective and the members have high motivation to accomplish the task, the more intelligent the members are, the better group performance will be. When the leader's style is directive or the members have low motivation, member intelligence will have little effect on group performance.*

**Table 4.** Correlations Between Member Intelligence and Group Performance

| | | Category | | |
|---|---|---|---|---|
| | *1* | *2* | *3* | *4* |
| | Nondirective Leader High Member Motivation | Nondirective Leader Low Member Motivation | Directive Leader High Member Motivation | Directive Leader Low Member Motivation |
| | n = 12 groups | n = 12 groups | n = 12 groups | n = 12 groups |
| | r = +.720*** | r = −.381 | r = +.014 | r = +.007 |

n    = Number of groups in category

r    = Correlation between member intelligence and group performance

*** = Statistically significant at .01 level (one-tail)

The findings in Category 1 of Table 4 show that, as predicted, high member intelligence has a fairly strong, positive influence on group performance when the leader is nondirective and the members are motivated. Also, as shown in Categories 2, 3, and 4, for groups with directive leaders or poorly motivated members, member intelligence has little effect on performance. For practical-use value comparison, of the 12 groups in Category 1, the groups with members of above-average intelligence had an average performance score in the top 25 percent of all 49 scores, while the groups with members of below-average intelligence had an average performance score at the 51st percentile. The average performance scores of the groups in Categories 2, 3, and 4, which had directive leaders or poorly motivated members, ranged between the 47th and 49th percentiles.

*In summary of this section, there are two key points to remember. First, if one is a nondirective leader and has enthusiastic subordinates, the more intelligent the members are, the better the unit will tend to perform. The reasons bright subordinates are more effective than less-intelligent subordinates are that the first can contribute better ideas and knowledge to the task planning and decisionmaking, and better planning decisions result in better accomplishment of the task. Second, if the leader is directive or his subordinates are poorly motivated, member intelligence has very little effect on performance.*

### Member Ability

As you probably already expect, the relationship between member ability and performance is very similar to that between member intelligence and performance. Remember that I defined ability to be the specific skills and task-relevant knowledge required during the performance phase to actually accomplish the work.

Although research studies disagree on whether member ability has an effect on performance, perhaps more for this relationship than any other in this book, there are everyday examples which clearly indicate that high member ability has a positive influence on performance. First, are not infantry platoons, submarine crews, or combat landing teams that have practiced together for a year better than similar-type units that have not practiced at all? Surely the reason the military puts so much time, effort, and resources into unit training is the belief that training increases people's ability and high ability improves unit performance. Second, if one takes 5 basketball players, 9 baseball players, or 11 football players of high talent or ability, will they not beat teams composed of poor quality players, other things being equal? Third, in the basic training environment, the new soldiers are required to perform as many pushups as they can on the physical readiness test. The reason that the average male trainee does more than twice as many pushups on the test than does the average female trainee is simply that the male is physiologically endowed with more ability, in this case upper body strength, than is the female.

It seems readily apparent that leadership style and member motivation will affect the actual amount of influence member ability has on the outcome. In the same manner the nondirective style allows member intelligence to affect group productivity, this leadership style permits member ability to influence unit performance. When the leader delegates authority to his subordinates and does not direct how every aspect of the work is to be done, success or failure depends in large measure on member skills. On the other hand, when the leader closely supervises and tells his subordinates what to do every step of the way, the outcome is due more to the leader's expertise and less to the members' talents. In the basic training battalion, I gave Foxtrot and Hotel a lot of free rein with how they conducted the training of their new soldiers because these two companies had excellent ability. The nondirective style let them use their own competence. My visits to their training sites were more to show them my interest in what they were doing than these visits were to check on the quality of training. With Golf, my method was exactly opposite. Golf had demonstrated that it was not very adept at training new soldiers, so I spent a great deal of time at their training sites, instructing and supervising the cadre as they trained the recruits. While using this directive style was very time consuming, it was my only effective option in that situation. As I had hoped, Golf Company's cadre eventually learned from the experience and became proficient enough that I was able to become considerably more nondirective with them. Thus, when the leader's style is directive, member ability will have little effect on performance. On the other hand, when the leader is nondirective, member abililty will have a fairly strong influence on the outcome.

The rationale for why member motivation ought to affect the influence of member ability on performance is basically the same as that presented earlier, when I discussed the relationship between member intelligence and productivity. Simply put, is it not logical that if one's subordinates are unwilling to do much more than show up for work and go through "the motions," then it doesn't really matter whether they have high or low ability? In this case, performance is not able to distinguish between those with talent and those without. The poor results will reflect only that a small amount of skill has been employed, and the actual ability level would have little bearing on the outcome. An example is the problem I faced with many drill sergeants during the last few months of their two-year tours of duty. These sergeants, who have the experience of training countless numbers of soldiers, had great ability to do their job. Yet, some of them began to "burn out" and made only a half-hearted effort because of the long duty hours, repetitive nature of the job, and considerable time spent away from their families. In those instances, performance was not a true reflection of their skills. On the other hand, when the members enthusiastically try the best they can, performance should accurately represent their ability.

High motivation, of course, does not ensure good performance. When ability is low, high motivation usually results in a lot of wasted effort. The new drill sergeant, although usually possessing good technical skills and high motivation, normally requires a few months' experience in order to gain the necessary teaching skills to become really effective. Once he has this expertise, high motivation produces excellent results.

I am suggesting that member ability will have little effect on group performance when the leader uses a directive style or the members are poorly motivated. On the other hand, when the leader's style is nondirective and the members are enthusiastic, high member ability will have a positive influence on performance and low member ability will have a negative influence on the outcome.

*Rule 5: When the leader's style is nondirective and the members have high motivation to accomplish the task, the more ability the members have, the better group performance will be. When the leader's style is directive or the members have low motivation, member ability will have little effect on group performance.*

The findings in Category 1 of Table 5 show that, as predicted, high member ability has a positive influence on group performance when the leader is nondirective and the members are motivated. Also, as shown in Categories 2, 3, and 4, for groups with directive leaders or poorly motivated members, member ability has little effect on performance. For practical-use value comparison, of the 12 groups in Category 1, the six groups with members of above-average ability had an average performance score in the top 25 percent of all 49 groups, while the six groups with members of below-average ability had an average performance score in the bottom 45 percent. The average performance scores of the groups in Categories 2, 3, and 4, which had directive leaders or poorly motivated members, ranged between the 47th and 49th percentiles.

*In summary of this section, there are two key points to remember. First, if one is a nondirective leader and has enthusiastic members, the higher the ability the members possess, the better the group will tend to perform. The reason those of higher ability are more effective than those of lower ability is that the first are able to shoot straighter, run faster, or do better at whatever skill the task requires. The difference between good and poor performance is, of course, no more than the difference in the ability the group displays. Second, if the leader is directive or his subordinates are poorly motivated, member ability has little influence on performance. A point the leader going into a new job might think*

**Table 5.** Correlations Between Member Ability and Group Performance

| | Category | | |
|---|---|---|---|
| *1* | *2* | *3* | *4* |
| Nondirective Leader High Member Motivation | Nondirective Leader Low Member Motivation | Directive Leader High Member Motivation | Directive Leader Low Member Motivation |
| n = 12 groups | n = 12 groups | n = 12 groups | n = 12 groups |
| r = +.559* | r = −.448 | r = −.476 | r = −.196 |

n = Number of groups in category

r = Correlation between member ability and group performance

* = Statistically significant at .05 level (one-tail)

*about is that if he doesn't know much about the work his new group performs, perhaps he ought to consider giving his subordinates, if they are talented, a good amount of leeway in how things are done. On the other hand, if his people don't have much ability, he'd better learn the required skills in a hurry and use the directive style.*

### Leader Intelligence

This and the next section address the influence that leader intelligence and ability have on performance. The rationale used to support the rules in these two sections is the same as I have used throughout. Namely, under the nondirective style, member intelligence and ability have a fairly strong influence on performance because the participative leader, who relies on both the advice of his subordinates and their competence to perform the work, will be successful to the extent his group members are knowledgeable and skilled. It follows that under the directive style, leader intelligence and ability will have a fairly strong influence on performance because this type of leader, who relies on his own expertise in planning and decisionmaking functions and who directs how the work is to be done, will be successful to the extent of his own competence. In this second case, the leader plays the major influencing role, while in the first case it is performed by the group members.

Although research studies disagree considerably on whether or not leader intelligence influences group performance, "real-world" selection criteria for leadership and executive positions leave no doubt that most organizations consider high leader intelligence to have a strong, positive effect on unit performance. The process used to select and train military officers is a clear illustration. With the single exception of the few who receive battlefield commissions, military officers earn their commissions by successfully completing programs with high intelligence requirements. For example, the Army's principal commissioning sources, the United States Military Academy, the Reserve Officers Training Corps program, and the Officer Candidate School, all require candidates to score well on mental entrance tests in order to be accepted for admission. Continuation in and graduation from these programs require further demonstrations of high mental competence. Those of low intelligence fall by the wayside. The Army's primary rationale for making these programs so intellectually demanding is that more-intelligent people will make better officers than will less-intelligent people because unit performance depends upon officer intelligence. The actions of student candidates and officers themselves clearly support my contention that the military services use this rationale. The strong competition to be among the honor graduates at the various precommissioning, junior, and intermediate-level service schools and the need that many officers feel to obtain a graduate degree both illustrate my point.

It is reasonable to expect leadership style and member motivation to have an effect on the influence that leader intelligence has on the outcome. Without restating the entire first paragraph of this section, unit performance under a directive leadership style must rely on the leader's talents. Conversely, under the nondirective style, performance is primarily a function of member competence. In the basic training battalion, I used the nondirective style with Foxtrot and Hotel, fully knowing that the talents of these two companies, rather than my own, would be the dominant influence on their performance. This did not bother me at all because these units were competent and had very sound ideas about how to train new soldiers. However, with Golf, I chose to be directive because I wanted to ensure that Golf's performance reflected my planning and decisionmaking skills rather than their own limited talents. Thus, when the leader's style is directive, leader intelligence will have a fairly strong influence on performance. When the leader is nondirective, his intelligence will have little effect.

Regarding member motivation, it seems essential that one's subordinates must wholeheartedly try to carry out their leader's plans and directives in order for leader intelligence to influence the outcome. What difference will the quality of the leader's planning make if his subordinates are unwilling to follow his guidance?

Even the directive leader must rely on his subordinates to some degree. He obviously cannot do all of the actual work by himself, nor can he supervise in more than one place at a time. Had I chosen to be directive with Echo Company during the period when its motivation was low, that company's productivity would not have increased. Thus, when group members merely go through the motions and do a minimum of work, the effect of the leader's intelligence on performance should be considerably reduced. On the other hand, when group members do their best to carry out their leader's guidance, performance ought to be influenced by the quality of the advice he gives them. In the basic training battalion, Golf Company's performance got steadily better because its cadre tried very hard to follow my guidance to the letter.

I am suggesting that leader intelligence will have little effect on group performance when the leader is nondirective or the members are poorly motivated. On the other hand, when the leader's style is directive and the members are enthusiastic, high leader intelligence will have a positive influence on performance and low leader intelligence will have a negative effect on the outcome.

*Rule 6: When the leader's style is directive and the members have high motivation to accomplish the task, the more intelligent the leader is, the better group performance will be. When the leader's style is nondirective or the members have low motivation, leader intelligence will have little effect on group performance.*

The findings in Category 1 of Table 6 show that, as predicted, high leader intelligence has a positive correlation with group performance when the leader is directive and the members are motivated. Also, as shown in Categories 2, 3, and 4, for groups with nondirective leaders or poorly motivated members, leader intelligence has little effect on performance. Of the 11 groups in Category 1, the groups with leaders of above-average intelligence had a considerably better average performance score than did the groups with leaders of below-average intelligence. Candidly, though, while the correlation in Category 1 (+.418) was positive and was the only correlation of appreciable size in the table, it did not reach statistical significance nor was it quite as high as I would have ex-

**Table 6.** Correlations Between Leader Intelligence and Group Performance

| | | Category | |
|---|---|---|---|
| *1* | *2* | *3* | *4* |
| Directive Leader High Member Motivation | Directive Leader Low Member Motivation | Nondirective Leader High Member Motivation | Nondirective Leader Low Member Motivation |
| n = 11 groups | n = 11 groups | n = 11 groups | n = 11 groups |
| r = +.418 | r = +.068 | r = −.043 | r = −.098 |

n = Number of groups in category

r = Correlation between leader intelligence and group performance pected. Simply put, this means that while there is reasonably good indication that Rule 6 is correct, the evidence in this case is not as strong as it is with the other relationships we have discussed.

Without belaboring the point too much, I must mention the circumstances. Analysis of the data shows that one of the 11 groups in Category 1 was a real maverick. For this particular group, the leader's intelligence score was high, but the unit performance score was the worst of all 11 groups in this category. A closer look shows that while this group reported they had above-average motivation, they also reported the lowest group cohesiveness score of all 49 groups. This large inconsistency suggests that this group's members probably didn't really work as hard as they reported they did and, thus, do not belong in Category 1. Although I must retain this group in Category 1 because to do otherwise would make suspect both the rules and the findings in this book, if this group was not included in the analysis, the correlation in Category 1 between leader intelligence and group performance would be both considerably higher and statistically significant.

*In summary of this section, there are two key points to remember. First, if one is a directive leader and has motivated subordinates, the more intelligent the leader is, the better his group will tend to perform. Bright leaders are more effective than less-intelligent leaders because the first contribute better ideas and knowledge to the task planning and decisionmaking, and better planning decisions result in better accomplishment of the task. Second, if the leader is nondirective or his subordinates are poorly motivated, leader intelligence has very little influence on performance. Thus, to effectively use the directive style, the leader must be bright and his subordinates enthusiastic.*

### Leader Ability

The relationship between leader ability and group performance is very similar to that between leader intelligence and performance. Recalling that ability consists of the specific skills and task-relevant knowledge required primarily during the performance phase, leader ability ought to have considerable influence on performance when the leader is directive and closely supervises and instructs his subordinates as they accomplish the work. On the other hand, when the leader is nondirective and delegates the authority to his subordinates regarding how they perform the work, leader ability should have very little effect on the outcome.

There are many everyday examples supporting the view that high leader ability has a positive influence on performance. A good illustration is the infantry company commander. During the performance or execution phase of whatever mission his unit undertakes, he uses his ability to oversee his subordinates' work by moving among his men, advising them, and correcting their mistakes. To do this well, he must obviously know how his subordinates are supposed to perform their individual tasks. Were this leader incompetent, unit performance would suffer because he would have few valuable suggestions to offer and would not be able to identify deficiencies in need of correction. It is rather easy to see the difference in performance between companies led by competent commanders and those led by the less competent. Capably led units cross the line of departure on time, know their location, use their assets efficiently, and have leaders at each command level who are on the scene directing activities when important things are happening. They fulfill both large and small requirements promptly and correctly. In units led by less-capable commanders, requirements, such as those mentioned above, are not performed nearly as well. The responsibility for unit performance clearly, and rightfully, rests with the unit commander.

A second example that high leader ability positively influences unit performance is that of the mess steward of a large dining facility. After he plans the menu, orders the staples, and assigns individual tasks to the cooks, the mess steward's job becomes supervisory. Unless he personally knows how to cook the meat and potatoes, serve the food, and keep the eating utensils and seating areas clean and available, how can he possibly ensure that his subordinates are doing their jobs right? This example of how the mess steward's ability affects dining facility performance strikes very close to home. During my first few months in the basic training battalion, I had several "opportunities" to explain to my brigade commander why my unit's dining facility had such poor inspection ratings. In my last discussion with him, I explained that my mess steward had been cited on several previous evaluation reports for incompetence. I also stated that my counseling sessions with the mess steward indicated he knew very little about the tasks his subordinates performed. My unit was rewarded by the assignment of an incoming mess steward with a very good reputation. To my mind, the fact that my unit's dining facility won the next eight consecutive awards for "Best Brigade Dining Facility of the Quarter" was not just coincidental with the new mess steward's arrival. Neither was the fact all 18 cooks in the dining facility scored over 90 percent (a rather unheard of occurrence) on the annual skill qualification test after they received training classes from the new mess steward.

It is apparent that both leadership style and member motivation will affect the influence that leader ability has on performance. When the leader is directive and closely supervises and instructs his subordinates as they perform their work, the outcome will depend upon the leader's skills and competence. On the other hand, when the leader is nondirective and leaves the actual performance of the work up to his subordinates' discretion, the outcome is based on the members' talents rather than on the leader's. Therefore, in the basic training battalion, my nondirective style with Foxtrot and Hotel caused my personal ability to have little effect upon how well these two companies trained their new soldiers. There was no problem with this choice of style because these units were competent in their own right. However, with Golf, I chose to be directive and supervise closely in order that my ability, rather than that company's limited talents, would be the dominant influence on soldier training.

Member motivation similarly affects the influence of leader ability. Even if the leader is directive and closely supervises his subordinates, if the group members choose to "fight him every step of the way" and only give the appearance of trying to do what he wants, his ability will have little effect on performance. This was the situation in which I found myself initially with Echo. That company was not only somewhat lazy, but its cadre members resisted, as much as they felt they could get away with, implementing some training changes I felt were necessary. I don't think Echo's cadre resented the "new guy" changing the way a few things were done nearly as much as they resented the fact that the changes entailed a little more work on everyone's part. Therefore, my ability initially had considerably less effect on improving the quality of soldier training in Echo than it did in Golf because Golf supported me with its best effort.

I am suggesting that leader ability will have little effect on group performance when the leader is nondirective or the members are poorly motivated. On the other hand, when the leader's style is directive and the members are enthusiastic, high leader ability will have a positive influence on peformance and low leader ability will have a negative effect on the outcome.

*Rule 7: When the leader's style is directive and the members have high motivation to accomplish the task, the more ability the leader has, the better group performance will be. When the leader's style is nondirective or the members have low motivation, leader ability will have little effect on group performance.*

The findings in Category 1 of Table 7 show that, as predicted, high leader ability has a fairly strong, positive influence on group performance when the leader is directive and the members are motivated. Also, as shown in Categories 2, 3, and 4, for groups with nondirective leaders or poorly motivated members, leader ability has little effect on performance. For practical-use value comparison, of the 11 groups in Category 1, the groups with leaders of above-average ability had an average performance score in the top 10 percent of all 49 scores, while the groups with leaders of below-average ability had an average performance score in the bottom 25 percent. The average performance scores of the groups in Categories 2, 3, and 4, which had nondirective leaders or poorly motivated members, ranged between the 39th and 59th percentiles.

**Table 7.** Correlations Between Leader Ability and Group Performance

| | *Category* | | |
|---|---|---|---|
| *1* | *2* | *3* | *4* |
| Directive Leader High Member Motivation | Directive Leader Low Member Motivation | Nondirective Leader High Member Motivation | Nondirective Leader Low Member Motivation |
| n = 11 groups | n = 11 groups | n = 11 groups | n = 11 groups |
| r = +.609* | r = −.150 | r = −.423 | r = +.102 |

n = Number of groups in category

r = Correlation between leader ability and group performance

* = Statistically significant at .05 level (one-tail)

*In summary of this section, there are two key points to remember. First, if one is a directive leader and has enthusiastic members, the greater the leader's ability, the better the group will tend to perform. The reason leaders with higher ability are more effective than those with lower ability is that the first are more knowledgeable concerning the skills required by the task and are, therefore, more competent as supervisors. Second, if the leader is nondirective or the members are poorly motivated, leader ability has little influence on performance.*

In chapter summary, the findings concerning the relationships between member and leader intelligence and ability and group performance lead us to three important conclusions. First, if one's subordinates are motivated to accomplish the task, member intelligence and ability have a strong effect on unit performance when the leader uses a nondirective style. In this situation, if member abilities are high, performance is good. If member abilities are low, performance is poor. When the leader chooses to use a directive style, member intelligence and ability have little influence on the outcome. Second, again given that one's subordinates are motivated to accomplish the task, leader intelligence and ability have a strong effect on unit performance when the leader uses a directive style. In this situation, if leader ability is high, performance is good. If leader ability is low, performance is poor. When the leader chooses to use a nondirective style, leader intelligence and ability have little influence on the outcome. It is readily apparent, then, that it is vitally important for the leader to use the style which will most effectively take advantage of the intelligence and ability within his group. Third, in all situations in which one's subordinates do not have sufficient motivation to accomplish the task, neither member nor leader intelligence nor ability have any considerable effect on performance. It follows, then, that if the leader devotes all of his time to improving the abilities of his subordinates and himself while at the same time he ignores the maintenance of unit motivation and morale, he is headed for trouble. *Captain Bligh of the HMS Bounty could tell us more about this.*

# 4

# Group Incentive: Member and Leader Motivation

The idea that high motivation has a positive influence on how well people will perform the task at hand probably enjoys more consensus of opinion than does any of the other relationships discussed in this book. Regardless of whether one looks at theory or at the emphasis those in leadership positions place on achieving high unit motivation, there is considerable agreement that "trying one's best" or failing to do so will make a considerable difference in how well the job is done.

### Member Motivation

Without getting into a long discussion concerning the several definitions of motivation available in the research, I consider a person's level of motivation simply to be the amount of effort he is willing to put into accomplishing the task. Those with high motivation work hard, while those with low motivation do not. My rationale for why enthusiastic units perform better than do poorly motivated groups, other things being equal, is very simply that high motivation uses a greater amount of the available talents and skills of the leader and the group members than does low motivation. Surely, if two units have equal skills and one of these groups uses its talents while the other does not, the first will achieve far better results than will the second.

"Real-world" examples clearly support the view that high member motivation has a positive influence on performance. Military history has countless cases of individuals and units whose high motivation to win and refusal to quit led either to victory or the best possible performance under the circumstances. Valley Forge, Pickett's Charge and the 20th Maine's defense at Gettysburg, Alvin York, Audie Murphy, and the Chosin Reservoir are but a few examples of how strong determination to succeed resulted in a better outcome than would have been achieved had the individuals involved made a lesser effort. In today's military, recalling proud unit histories and the dignity and splendor of ceremonies and traditions, among other things, serve to inspire servicemembers to do their duty to the best of their ability. The services firmly believe that a great deal of the difference between victory and defeat lies in whether the unit members are willing to "give it their all."

It seems apparent that member ability and group cohesiveness should affect the actual amount of influence that member motivation has on the outcome. Regarding member ability, it would seem necessary for the group members to have a certain amount of skill in order for motivation to improve performance. In other words, it is difficult to understand how a high level of motivation would improve the productivity of people who cannot adequately perform their work. I have seen countless examples of men and women who wanted very much to be in the Army and who tried very hard to complete successfully the requirements of basic training but failed because of reading disabilities, uncorrectable physical impairments, or mental deficiencies. These trainees just did not have the ability to take advantage of their high motivation. On the other hand, when group members are talented, the level of their motivation ought to have considerable effect on performance. Those groups who use their talents will perform well, while those who do not will perform poorly. In the basic training battalion, while Echo, Foxtrot, and Hotel were each very competent, only the last two companies initially had high motivation. It was not a coincidence that during the first few months, Foxtrot's and Hotel's performances far outshone that of Echo. Thus, when one's subordinates are incompetent, member motivation will have little effect on performance. When one's subordinates are talented, however, member motivation will have a fairly strong influence on the outcome.

It seems reasonable that the group's cohesiveness ought to affect the influence of motivation on performance. Cohesiveness, of course, reflects the extent to which the unit members display cooperative support, mutual assistance, organizational coordination, and teamwork in accomplishing the common task. My rationale is that when the unit members have a good working relationship among themselves, the unit more effectively and efficiently utilizes its ability and motivation. It follows, then, that member motivation would have considerably more influence on the outcome if the group was cohesive than if it was not. Specifically, without a unified and cohesive effort, some of each individual's efforts will be wasted in duplicating work, in gathering information that other members already have, or in pursuing the wrong direction. Thus, when group members are uncohesive and withhold assistance, cooperation, and information from each other for personal advancement, spite, or whatever reason, member motivation should have considerably less effect on performance than it would in a coordinated, cohesive venture. On the other hand, when members are cohesive, their motivation will have a fairly strong effect on performance. The efficiency with which these cohesive members channel their motivation into task accomplishment causes their efforts to have a fairly strong infuence on the outcome. Further, it stands to reason that highly motivated, cohesive groups will perform well, while poorly motivated, cohesive units will not.

In the basic training battalion, although Echo, Foxtrot, and Hotel were all cohesive units, only Foxtrot and Hotel initially had high motivation. These two companies far outperformed Echo. The reader may wonder how Echo could have been very cohesive yet poorly motivated to accomplish the task. The answer is that the high cohesion among Echo's cadre unfortunately took the form of sharing information and providing assistance to each other mostly in regard to hunting, fishing, and getting along well, rather than training soldiers.

I am suggesting that member motivation will have little effect on group performance when the members have low ability or the group has low cohesiveness. On the other hand, when the group members are talented and the group is cohesive, high member motivation will have a positive influence on performance and low member motivation will have a negative effect on the outcome.

*Rule 8: When the members have high ability and the group has high cohesion, the more motivation the members have, the better group performance will be. When the members have low ability or the group has low cohesion, member motivation will have little effect on group performance.*

The findings in Category 1 of Table 8 show that, as predicted, high member motivation has a fairly strong, positive influence on group performance when the members have high ability and the group has high cohesion. Also, as shown in Categories 2, 3, and 4, for groups with members of low ability or low cohesion, member motivation has little effect on performance. For practical-use value comparison, of the 12 groups in Category 1, the 7 groups with members of above-average motivation had an average performance score in the top 25 percent of all 49 groups, while the 5 groups with members of below-average motivation had an average performance score in the bottom 37 percent. The average performance scores of the groups in Categories 2, 3, and 4, which had low ability members or low group cohesion, ranged between a low of the 25th percentile to a high of the 61st percentile.

Before summarizing the findings in this section, there is one additional area I ought to address. Establishing the conditions under which member motivation has a positive effect on performance solves only half of the reader's problem. The other half is, of course, identifying the factors that cause member motivation to increase or decrease. To examine this second problem, I computed correlations between member motivation and every other factor and behavior that I measured during this study. The findings show that, first, although there was little correlation between member

**Table 8.** Correlations Between Member Motivation and Group Performance

*Category*

| 1 | 2 | 3 | 4 |
|---|---|---|---|
| High Member Ability High Group Cohesion | High Member Ability Low Group Cohesion | Low Member Ability High Group Cohesion | Low Member Ability Low Group Cohesion |
| n = 12 groups | n = 10 groups | n = 10 groups | n = 13 groups |
| r = +.619** | r = −.161 | r = −.018 | r = −.253 |

n   = Number of groups in category
r   = Correlation between member motivation and group performance
** = Statistically significant at .025 level (one-tail)

motivation and either leader intelligence or leader ability, there was a fairly strong, positive correlation ($r = +.552$****, $n = 49$ groups) between member motivation and how well the members felt their leader was doing his job. In other words, in groups where the members felt their leader was doing a good job, their motivation was high. In groups where the members felt their leader was doing a poor job, their motivation was low. It makes good sense that members will feel more motivated when working for someone they believe is competent than when they are working for someone they think doesn't know what he is doing. When subordinates feel their leader is capable, they believe his talents will ensure that their hard work pays off in good unit performance. On the other hand, when subordinates feel their leader is incompetent, they believe that unit performance will be poor despite their hard work. Thus, to achieve high member motivation, it is important that the leader be competent and know how to get the job done right. If he doesn't have these abilities, his subordinates will think poorly of him and just won't put as much effort into their work.

A second finding was that while member motivation had virtually no correlation with member ability, it had a fairly strong, negative correlation ($r = -.531$****, $n = 49$ groups) with member intelligence. In other words, the relatively bright members had lower motivation than did the less talented people. I believe this finding is only applicable in certain situations rather than in all cases. The subjects of this study were cooks assigned to work in Army dining facilities. Their job is certainly important to their units, but it is often thought of, by the cooks themselves, as thankless and of little prestige. That people who fail to qualify for other Army job specialties sometimes end up assigned as cooks adds to their poor self-image. Therefore, the apparent cause of this negative correlation is that the cooks of relatively higher intelligence were rather disgruntled with their lot in life, while those of relatively lower intelligence were happy to be gainfully employed. This example points out two things the leader needs to be concerned with. First is the problem of overqualification. When possible, the leader should assign people to duties which challenge their abilities and are of interest to them. Second, the leader needs to remember to talk to his subordinates, telling them that they are doing a good job and that their work and efforts are important to the unit's success.

A third finding was that while member motivation had little correlation with whether the leader's style was directive or nondirective, there was a fairly strong, positive correlation (r = +.513****, n = 49 groups) between member motivation and whether the members felt their leader would listen to and consider implementing their suggestions and recommendations. In other words, although it apparently did not affect the cooks' motivation whether or not the mess steward reserved the decisionmaking authority for himself, it did matter to the cooks that they at least had the opportunity to offer their ideas. When the leader asked for their opinions, their motivation was high. But when the cooks felt they had no say in how things were done, their motivation suffered. The subtle point to remember is that the leader does not have to follow the advice of his subordinates, but he would be wise to ask them for their opinions. If their suggestions and recommendations are good, the leader should put these ideas into practice and give the members the credit for them. Subordinates will solidly back the leader's plan when they contribute to it. If, however, the advice from one's subordinates is poor, the leader should tactfully point out the flaws in that advice and go a better route. If he does, not only will his subordinates gain confidence in their leader's ability and judgment, but they will not become less motivated because he did not follow their well-intentioned advice.

A fourth finding was that member motivation had a fairly strong, positive correlation (r = +.565****, n = 49 groups) with leader motivation. This finding clearly shows why correlation does not, by itself, prove causality. Is high leader motivation causing high member motivation or is the reverse true? I believe the answer in this case is that each factor has some effect on the other. On the one hand, when leaders give their best effort, they inspire their subordinates to work hard, while when leaders show disinterest in the task, they cause a lack of concern among the group members. Thus, the familiar Army saying, "the unit does well on the things the leader is concerned with." On the other hand, when one's subordinates are "giving it all they have," they are a great source of pride and inspiration to the leader, while when the leader just can't get his subordinates to make the effort they are capable of, his motivation suffers. Therefore, leader and member motivation each serve to affect the other. From the leader's standpoint, he should remember that when he is willing to work the long hours, share the burdens of the weather, and maintain high enthusiasm for the task, his motivation will rub off on his subordinates and cause them to work hard.

A fifth finding, which may surprise some, is of considerable importance to unit readiness. The data showed that member motivation had a positive correlation ($r = +.450****$, $n = 49$ groups) with leader enforcement of high performance standards. In the past, there have, unfortunately, been quite a few leaders who believed that "taking it easy" on their subordinates or "being one of the boys" would cause group members to have high motivation and give a good effort. Indeed, the beginnings of the present all-volunteer Army, when "the Army wants to join you" was in vogue and the discipline of subordinates was relaxed, is a clear example of this belief. The fact of the matter is that the data show that subordinates want to be part of a disciplined unit in which members are held accountable for their performance. Subordinates are not dumb. They know that if some members can get away with doing nothing, unit performance will suffer and the remaining members will have to do more than their share of the work. Also, undisciplined units have more interpersonal problems, crime, and the like. It's just no fun being part of a mob. On the other hand, the findings in Chapter 2 showed that leader enforcement of high standards causes unit performance to be good. Subordinates have pride in belonging to a good unit. People have a sense of fairness and equity when those who do well are rewarded and those who break the rules or make no effort are punished. Leaders need to remember that subordinates understand that discipline and hard work are necessary for success and these people joined the organization in order to be successful and to be proud of themselves. Thus, when the members see the leader enforce high performance standards, they work hard because they know they will get a fair return for their effort.

The behavior questionnaire I used in this study was designed to determine what influence member motivation, among other things, has on group performance. My main intent was not to determine which factors cause motivation to increase or decrease, so the questionnaire did not survey all of the factors which could possibly influence member motivation. It follows, then, that while this book is able to show some of the factors which affect member motivation, I do not consider the factors presented here to be an all-inclusive list.

Finally, I need to discuss one more point concerning member motivation because I had a problem with it. At the beginning of my second basic training cycle, I actually did institute in the battalion the "Best Company of the Cycle" award that I mentioned earlier. The basic advantage of such an award, it is hoped, is that it causes one's subordinates to work hard. This was clearly the result. The obvious disadvantage is that the competition may get so intense that the companies stop helping each other. If the situation gets out of hand, it can lead to cheating or similar activities in order to get "good statistics." During my third cycle, it seemed the companies were not sharing ideas as easily as they had previously. Sensing that competition might be the cause of it, I made it a point at my weekly planning meetings with the company commanders to discuss the good training ideas I had observed that week while visiting the individual company training sites. To put it simply, if the companies were not going to share their good ideas, I was going to spread the word myself. During my fourth cycle, it seemed the companies were working longer than their already hard schedule required, the cadre's tempers were getting shorter with the new soldiers, and the companies were losing their rapport with each other. At that point, I changed the award system.

While the old system had only one overall winner, the new system gave awards for each major training subject (rifle marksmanship, physical readiness, drill and ceremonies, and soldier skills). Further, inter-company competition was done away with.

Under the new system, every company that met pre-designated performance standards would win streamers they could attach to their company guidon. Thus, all four companies could win up to four awards apiece, regardless of how well the other companies performed. Soon, the companies were working better together than they had originally and the overall battalion performance improved considerably. The important point here is that giving awards in order to motivate subordinates to achieve high performance is much more effective when the awards are based upon achieving certain performance standards as opposed to out-performing other units.

*In summary of this section, there are three key points to remember. First, when the group has high member ability and high cohesion, the higher the level of motivation the members possess, the better the group will tend to perform. The reason those with high motivation perform better than do those with low motivation is simply that the first are willing to use more of their abilities, both quantitatively and qualitatively, and performance is just the measure of these expended abilities. Second, if the group has low member ability or low cohesion, member motivation has little influence on performance. Third, there are several factors that will cause member motivation to increase. Among these are member perceptions that their leader does his job well, the members themselves are gainfully employed and contributing to their unit's success, their leader has an interest in the members' opinions of how things should be done, their leader is highly motivated to accomplish the task, and their leader enforces high performance standards.*

### Leader Motivation

With the great amount of study and writing that researchers have devoted to the field of motivation, they have done surprisingly little work on the sub-topic of leader motivation. Page after page presents theories of the relationships between things such as needs, incentives, and levels of satisfaction. Additional pages describe factors that cause subordinates to be more or less motivated. Still other pages discuss concepts such as the idea that subordinates are not all motivated equally by the same incentives. When the group leader is mentioned, it is usually to point out actions he can take to increase the motivation of the people who work for him. Yet, surely, the degree to which the leader himself is motivated to ensure his unit performs the task as well as possible must have an effect on the outcome. Is it not reasonable to believe that a good deal of the considerable success achieved by the US Third Army during World War II was because of the great personal motivation and drive of its commander, General George S. Patton, Jr., to have "the best Army ever put in the field"? Does not General of the Army Douglas MacArthur's quotation concerning the value of rigorous athletic training to the character development of West Point cadets

> Upon the fields of friendly strife
> Are sown the seeds that,
> Upon other fields, on other days,
> Will bear the fruits of victory

mean that the resolve of leaders to "give it their all" and their refusal to quit are critical to their units' success or failure?

Perhaps one reason we find so little research concerning the effect of leader motivation on group performance is because the relationship between these two factors is rather complex. First, leader motivation has a direct influence on performance, but only in certain situations, and second, leader motivation also has an indirect effect on the outcome by influencing other factors which themselves cause performance to increase or decrease. Let me explain.

It seems reasonable that the leader's ability and whether or not he establishes high performance standards will affect the amount of direct influence that leader motivation has on the outcome. Regarding ability, one action the motivated leader can take to improve unit performance is to use his personal knowledge of work methods and techniques to advise and guide his subordinates in the performance of their work. Of course, to be successful, the leader must have high ability. If the leader has little talent, how can it possibly matter whether or not he is motivated to give advice to his group members? Certainly, giving poor advice to one's subordinates is of no more value in terms of performance than is giving no advice at all. On the other hand, when the leader is competent and knowledgeable, if he is motivated to help his group members by offering them good ideas and suggestions, particularly when they are of below-average ability, performance should improve. This is clearly the situation that existed with Golf. That company, despite its best efforts, was not doing well. My suggestions to the cadre, regarding better ways to train the new soldiers, helped to improve that unit's performance. But if the competent commander, through a lack of motivation, chooses to let his subordinates "go it alone," he does nothing to make the outcome any better. In other words, had I failed to try to help Golf, its performance would have remained poor. Thus, when the leader has little ability, his motivation will have little effect on performance. When the leader is competent, his level of motivation will have a fairly strong influence on the outcome.

The leader can also improve his group's performance by establishing high performance standards. Remember that enforcement of high standards consists of both establishing an above-average minimum level of acceptable performance for the group members and displaying above-average insistence that those subordinates meet the defined level. The leader who takes only one of these actions, or neither, is not enforcing high standards. As an example, if the leader establishes the admirable requirement that 100 percent of his unit members pass the periodic physical readiness test and then neither requires any remedial training nor imposes any sanctions against those who fail the test, he is obviously not enforcing high standards. Neither is the leader who ensures that all of his unit members meet a particular requirement by establishing such a low performance standard that it takes his subordinates little effort or skill to pass. It is readily apparent, then, that the level to which the leader establishes high standards should be a factor in determining the amount of influence leader motivation has in any specific situation. In the basic training battalion, if the commander establishes a low minimum level of acceptable performance, how can it matter whether or not he is motivated to ensure the group members meet that level? Performance will be low and not appreciably different in either case.

When the leader establishes high minimum levels of acceptable performance, however, the motivated leader will achieve excellent results by insisting that the members meet those standards, while the unmotivated leader will achieve poor results by failing to insist that his subordinates perform to the required level. This is the situation that existed with Echo as the battalion practiced for the Command Retreat ceremony mentioned earlier. Clearly, I had established a high level of expected performance and, equally clearly, Echo was not motivated enough to reach that required level. My high motivation to have that company perform well, which prompted my conversation with the cadre members regarding their duties and responsibilities, caused that unit to perform well during the ceremony. If I had established a high level of expected performance and, then, because of a lack of motivation on my part, failed to insist that Echo meet the standard, that company's performance during the ceremony would have been poor. Therefore, when the leader fails to establish high standards, his motivation will have little effect on performance. When the leader does establish high standards, however, his level of motivation to insist that the established standards be met will have a fairly strong influence on the outcome.

I am suggesting that leader motivation will have little effect on group performance when the leader has low ability or fails to establish high performance standards. On the other hand, when the leader is competent and establishes high standards, high leader motivation will have a positive influence on performance and low leader motivation will have a negative effect on the outcome.

*Rule 9: When the leader has high ability and establishes high performance standards, the more motivation the leader has, the better group performance will be. When the leader has low ability or fails to establish high standards, leader motivation will have little effect on group performance.*

The findings in Category 1 of Table 9 show that, as predicted, high leader motivation has a fairly strong, positive influence on group performance when the leader has high ability and establishes high standards. Also, as shown in Categories 2, 3, and 4, when the leader has low ability or fails to establish high standards,

**Table 9.** Correlations Between Leader Motivation and Group Performance

| | Category | | |
|---|---|---|---|
| *1* | *2* | *3* | *4* |
| High Leader Ability High Established Standards | High Leader Ability Low Established Standards | Low Leader Ability High Established Standards | Low Leader Ability Low Established Standards |
| n = 12 groups | n = 11 groups | n = 12 groups | n = 9 groups |
| r = +.568* | r = −.125 | r = +.185 | r = +.404 |

n = Number of groups in category
r = Correlation between leader motivation and group performance
* = Statistically significant at .05 level (one-tail)

leader motivation has little effect on performance. The positive correlation ( + .404) in Category 4, which is somewhat higher than expected, does not closely approach statistical significance levels and appears to be a matter of chance because it seems theoretically unsupportable. For practical-use value comparison, of the 12 groups in Category 1, the 6 groups with leaders of above-average motivation had an average performance score in the top 33 percent of all 49 groups, while the 6 groups with leaders of below-average motivation had an average performance score in the bottom 25 percent. The average performance scores of the groups in Categories 2, 3, and 4, which had leaders of low ability or who failed to establish high standards, ranged between a low of the 33rd percentile to a high of 50th percentile.

As mentioned earlier, leader motivation also affects the outcome indirectly by influencing other factors which themselves cause performance to be better or worse. Specifically, leader motivation correlated positively with leader enforcement of performance standards ($r = +.709^{****}$, n = 49 groups), member motivation ($r = +.565^{****}$, n = 49 groups), and group cohesiveness ($r = +.491^{****}$, n = 49 groups). In other words, as leader motivation increased or decreased, each of these three factors tended to move in the same direction. Obviously, raising the level of these three factors does not improve performance in every situation. But, as seen in Rules 3 and 10, in groups with competent and enthusiastic members, raising the level of leader enforcement of standards and group cohesiveness clearly improves performance. Further, as pointed out in Rule 8, in groups with competent and cohesive members, raising the level of member motivation clearly improves performance. Thus, raising the level of leader motivation tends to raise the levels of leader enforcement of performance standards, member motivation, and group cohesion which, in turn, improves unit performance in certain situations. Of course, I need to address the issue of causality for these three correlations.

First, regarding the correlation between leader motivation and enforcement of high standards, it is reasonable that the level to which the leader is motivated will influence the level to which he enforces high standards, rather than the reverse being true. Did I not require Echo and Golf to practice the Command Retreat ceremony over and over because I was motivated to have the battalion perform well? It just makes no sense to say the opposite, that it was the repeated practice that caused me to be motivated.

Second, concerning the correlation between leader and member motivation, I pointed out earlier why each of these factors has an effect on the other. In brief, when the leader works hard, he is an inspiration to his subordinates, but when the leader is disinterested in the task, he will cause a lack of effort among his group members. On the other hand, when one's subordinates try their best, they are a source of pride and inspiration to the leader, while when the leader just can't get his subordinates to make the effort they are capable of, his motivation suffers. Certainly, then, while member motivation has an influence on the leader, there is a clear rationale for why the leader's motivation has an effect on his subordinates' efforts.

Third, regarding the correlation between leader motivation and group cohesiveness, it is reasonable that the level to which the leader is motivated will influence the level to which the group is cohesive. My rationale is based on the positive correlation which exists between member motivation and group cohesiveness. In other words, high member motivation tends to increase the cooperative support, mutual assistance, organizational coordination, and teamwork among the group members. It follows, then, that because high leader motivation improves the members' motivation and high member motivation improves group cohesiveness, the extent to which the leader works hard will affect the cohesiveness of his group.

To summarize, high leader motivation serves, indirectly, to improve unit performance by raising the levels of leader enforcement of standards, member motivation, and group cohesiveness which, in their own right, positively influence the outcome. Conversely, using the same rationale, low leader motivation serves to decrease performance.

In the preceding section, I identified several factors which cause member motivation to improve. Similarly, it is important here to discuss the factors which cause leader motivation to increase. Although the study questionnaire was not designed to inquire about which factors serve to motivate the leader, certain things seem apparent from the available information.

First, one often overlooked fact is that the leader, at whatever organizational level, is also a member of another group. For example, the company commander, while being the leader of his own unit, is also a member of the group of company commanders and staff officers led by the battalion commander. In his role as a member of that group, the company commander is, of course, motivated by the same things that serve to motivate the people who work for him. Thus, the amount of effort the company commander makes is affected by his perception of whether the battalion commander does his own job well, asks for the opinions of the company commanders, is motivated to do a good job, enforces high standards for the battalion, and assigns important and worthwhile tasks to that particular company commander.

Second, as explained earlier, leader motivation is influenced by the amount of effort his subordinates are willing to make. Clearly, in the basic training battalion, it "did my heart good" to visit the training sites of Foxtrot, Golf, and Hotel because these units were always doing the best they could. On the other hand, when I visited Echo, I often departed their training site frustrated by the thought that they could do so much better if only they would try harder.

Third, although the questionnaire did not measure leader opinions about job satisfaction or the opportunity for advancement when one's unit performs well, these two factors obviously matter to some extent to most leaders. Experience indicates it is important to the unit's success and well-being that the first factor, trying to do a good job for its own sake, is of more concern to the leader than is the second, trying to do a good job in order to get ahead.

Finally, while the factors mentioned in this section clearly have an effect on leader motivation, they are not intended to be considered as an all-inclusive list of leader motivators.

*In summary of this section, there are four key points to remember. First, when the leader is competent and establishes high performance standards, the higher the leader's motivation, the better the group will tend to perform. The basic reasons that the highly motivated leader is more effective than is the poorly motivated leader are that the first contributes more to the task, in the form of good ideas and suggestions, than does the second, and the first requires more effort, both quantitatively and qualitatively, from the group members during the performance phase. Second, if the leader has low ability or does not establish high standards, leader motivation has little effect on the outcome. Third, high leader motivation has an additional positive influence on performance, indirectly, because it raises the levels of leader enforcement of high standards, member motivation, and group cohesion which, by themselves, have a positive influence on performance. Fourth, there are several factors which will cause leader motivation to increase. Among these are high subordinate motivation, high leader job satisfaction, and the expectation that the leader will advance if his unit performs well. Other motivating factors include the leader's perceptions of whether or not his boss does his job well, asks for his subordinates' opinions regarding how things should be done, is motivated to accomplish the task, enforces high performance standards, and assigns worthwhile and important tasks to his subordinate leaders.*

# 5
# Group Bonding: Unit Cohesion

During the past decade, the military services have devoted a considerable amount of attention and study to group or unit cohesion. One reason for this interest is that as the military strives to constantly improve its readiness posture in a period when fiscal resources are relatively limited and the United States' strongest military adversary holds quantitative advantages in manpower strength and equipment, there is a search to find and use any factor which will serve as a "combat force multiplier." Certainly, military history supports the belief that unit cohesion is such a multiplier. Another reason for the increased interest in cohesion is that a number of senior military leaders and analysts believe that during the latter stages of the Vietnam war and the initial few years of the All-Volunteer Force, low unit cohesion was a significant contributor to a perceived decrease in unit motivation, discipline, and performance.

Some analysts suggest that unit cohesion declined because the military has shifted away from the personnel-oriented leadership approach toward the more impersonal management approach which focuses on the allocation of time and resources, and recruiting servicemembers based on pay incentives creates a commitment more to oneself than to the organization. For those interested in more detail, Johns et al. (1984) present a detailed explanation of these processes.

## Cohesion

While most people agree that high group cohesiveness has a positive influence on unit performance, there is considerably less agreement as to exactly what group cohesion is. Definitions range from the concept that cohesion is "the resultant of all the forces acting on all the members to remain in the group" to the description that "a group is cohesive if the members feel attracted to the group or if the members are adjusted to the group" to the idea that "cohesion refers to a condition that causes members of a group to conform to certain standards of behavior and to subordinate self-interest to that of the group." I believe these definitions are inadequate because they fail to pinpoint the member interactions within the unit that serve to increase the unit's performance. As stated in Chapter 4, I define unit cohesion as the extent to which the members display cooperative support, mutual assistance, organizational coordination, and teamwork in accomplishing the common task. Units with high cohesion have these qualities, while units with low cohesion do not.

My rationale for why cohesive units perform better than do uncohesive groups, other things being equal, is that the first, with their good working relationships among the members, more efficiently use group assets such as ability, time, and equipment. In the basic training battalion, it makes good sense that if the four companies work well together by pooling their ideas and knowledge; dividing the task so that each company works on the area in which it has the most expertise; loaning each other equipment in short supply; and establishing good coordination regarding milestones, priorities, and the like, the battalion will perform much better than if the companies refused to cooperate with each other. As explained in the preceding chapter, this is exactly what happened once I changed the battalion award system from a one-winner, inter-company competition concept to a system in which all companies scoring higher than a certain performance standard won awards. The very noticeable increase in cohesion among the units was accompanied by significantly better average performance scores in each of the four major training areas.

Another example that high group cohesiveness has a positive influence on group performance involves the level of cooperation among the new soldiers in basic training. The platoons that had high trainee cohesiveness almost always performed better on rifle marksmanship, physical readiness, drill and ceremonies, and soldier skill tests than did platoons with low cohesiveness because in the first groups, the more talented soldiers voluntarily spent much of their free time teaching and coaching the less talented soldiers. In the uncohesive platoons, there was very little peer assistance and coaching.

It seems reasonable that member ability and member motivation should affect the amount of influence that group cohesion has on the outcome. Regarding member ability, it would seem necessary for the group members to have a certain amount of skill in order for cohesion to improve performance. In other words, if the members have little talent, performance will be relatively poor regardless of whether or not they are cohesive and work well together. In the basic training platoons just mentioned, it surely would not do performance much good if, in the cohesive platoons, the less talented soldiers were coached and assisted by fellow soldiers of equally poor skills. That is just a case of "the blind leading the blind." On the other hand, when the members are talented, the group's cohesiveness should affect performance. Members who receive useful ideas, assistance, and support from competent co-workers should achieve good results, while members who work in an uncohesive group, where the attitude is "everyone for himself," will not receive any help and, thus, their performance will not profit from their co-workers' high competence. Therefore, when the group members have low ability, cohesiveness will have little effect on performance. But when one's subordinates are talented, cohesiveness will have a fairly strong influence on the outcome.

Regarding member motivation, should not the extent to which the members are willing to work hard affect the influence cohesion has on the outcome? Surely, even if the members are competent to perform their work, if they are not willing to make a good effort, it matters little whether they are cohesive or not. In the basic training battalion, this was my initial situation with Echo. Despite the fact that the company was cohesive, that unit's failure to put forth the necessary effort caused it to perform poorly. Had Echo been uncohesive in this situation, the outcome would not have been perceptibly different. On the other hand, when one's subordinates are motivated to accomplish the task, cohesive groups will perform better than uncohesive groups simply because the first will use group assets more efficiently. Remember the example of how the battalion performed so much better after the award system was changed from companies competing against each other to companies competing against a performance standard? Before the change, the four companies were not cohesive, while after the change, they worked well together. Although the companies gave their best efforts under both award systems, their performance was much more effective when they worked cohesively.

I am suggesting that group cohesion will have little effect on performance when the members have little talent or are poorly motivated to accomplish the task. On the other hand, when the members are competent and enthusiastic, high group cohesion will have a positive influence on performance and low group cohesion will have a negative effect on the outcome.

*Rule 10: When the members have high ability and high motivation to accomplish the task, the more cohesion the group has, the better group performance will be. When the members have low ability or low motivation, group cohesion will have little effect on group performance.*

The findings in Category 1 of Table 10 show that, as predicted, high group cohesion has a fairly strong, positive influence on group performance when the members have high ability and high motivation. Also, as shown in Categories 2, 3, and 4, for groups with members of little talent or poor motivation, group cohesion has little effect on performance. For practical-use value comparison, of the 12 groups in Category 1, the 7 groups with above-average cohesion had an average performance score in the top 25 percent of all 49 groups, while the 5 groups with below-average cohesion had an average performance score in the bottom 16 percent. The average performance scores of the groups in Cate-

**Table 10.** Correlations Between Group Cohesion and Group Performance

| | Category | | |
|---|---|---|---|
| *1* | *2* | *3* | *4* |
| High Member Ability High Member Motivation | High Member Ability Low Member Motivation | Low Member Ability High Member Motivation | Low Member Ability Low Member Motivation |
| n = 12 groups | n = 10 groups | n = 13 groups | n = 10 groups |
| r = +.692** | r = +.091 | r = +.071 | r = −.518 |

n = Number of groups in category
r = Correlation between group cohesion and group performance
** = Statistically significant at .025 level (one-tail)

gories 2, 3, and 4, which had members of low ability or poor motivation, ranged between a low of the 35th percentile to a high of the 58th percentile.

It is important to identify and discuss those factors which cause group cohesion to increase or decrease. As was the case with member and leader motivation, I computed correlations between group cohesion and every other factor and behavior that I measured during this study. The most obvious finding was that the factors that had a fairly strong influence on group cohesion were the same factors, with one exception, that had a fairly strong influence on member motivation.

First, group cohesion correlated positively ($r = +.541{****}$, $n = 49$ groups) with leader enforcement of standards. Regarding causality, it seems much more reasonable that enforcement of standards will influence group cohesion rather than the opposite. It just doesn't make much sense to say that the more the group is cohesive, the more the leader tends to enforce high standards. There is, however, a clear rationale for why the enforcement of high standards increases group cohesion. The reasons are that when the leader establishes a high level of expected performance and insists that the members meet this level, he causes the members to "pull together" in order to reach the objective and causes the unit to perform well. The bonds formed between people when they have to depend upon each other in order to achieve a common goal considerably increase unit cohesion. So does the shared feeling of being part of a unit that does well. Thus, when the leader enforces high performance standards, the group tends to be much more cohesive than if the leader does not enforce these standards.

Second, group cohesion had a fairly strong, positive correlation ($r = +.511{****}$, $n = 49$ groups) with the members' perception of how well their leader was doing his job. When the members felt their boss was doing his job well, group cohesion was high. In groups where the members felt their leader was doing a poor job, cohesion was low. As one might guess, when members feel their boss is competent, the leader and his subordinates will have a good working relationship because the members believe that the leader's talents will ensure good unit performance. As mentioned earlier, people want to be part of a winning unit. On the other hand, when the members feel their boss is incompetent, unit cohesion will be lower because the members lack confidence in their leader's ability to do his job right and to produce good performance. This feeling will erode the working relationship between them and their boss.

Third, group cohesion correlated positively (r = +.491****, n = 49 groups) with leader motivation. As discussed in the last chapter, the reasons leader motivation influences cohesion are that high leader motivation causes an increase in member motivation because the leader who works hard inspires his subordinates to do likewise; and this high member motivation generally increases group cohesiveness because these members, who want to do a good job, realize that if they cooperate and assist each other, performance will be better than if they do not cooperate. Thus, when the leader works hard, group cohesion will be better than if the leader does not make a good effort.

Fourth, group cohesion had a positive correlation (r = +.482****, n = 49 groups) with how much the leader asked the group members for their opinions about how things should be done. When the leader actively sought his subordinates' ideas, group cohesion was high. When the members felt they had no say in things, group cohesion was low. Certainly, when the leader does not allow the members the opportunity to make suggestions, the working relationship between them is considerably more strained than when the members are able to contribute. In the second case, the joint participation of the leader with his subordinates creates a bond of teamwork and rapport that improves unit cohesion. Thus, when the members are able to contribute their suggestions to the task, group cohesion is better than when the members do not have the opportunity to contribute.

Finally, because the study questionnaire was not designed to determine which factors serve to increase group cohesion, the four factors identified here are not meant to be considered as the only factors which could influence cohesion.

*In summary of this section, there are three key points to re-member. First, when the group members are competent and enthu-siastic, the higher the level of group cohesion, the better the group will tend to perform. The reason that groups with high cohesion are more effective than those with low cohesion is that the first, with their good working relationships among the members, more efficiently use group assets such as ability, motivation, and time. Second, when the group has members of poor talent or low motiv-ation, group cohesion has little influence on performance. Third, there are several factors that will cause group cohesion to in-crease. Among these are member perceptions that the leader en-forces high performance standards, does his job well, is motivated to accomplish the task, and asks his subordinates for their opin-ions concerning the work.*

### Endnote: The Ability Factor

As an endnote to Rules 1–10, I need to address one final point. In each of the 10 rules or relationships presented in this book, the effect that the principal factor had on group performance was influenced by the presence or absence of two other factors. For example, the effect of member motivation on performance was influenced by the levels of member ability and group cohesion. The point of this endnote is that we must also consider, under lim-ited circumstances, a third influencing factor: the presence or ab-sence of a minimum level of member ability to perform the task. Certainly, it seems reasonable that we ought to consider the level of member ability, at least to some extent, in all 10 rules. After all, in every case, the members are the ones who must actually perform the work. The consideration of member ability as a third influencing factor is limited to those relationships in which mem-ber ability is not the principal factor affecting performance or one of the two included influencing factors. Thus, in the relationship between member motivation and performance, member ability, as a third influencing factor, is of no concern because it is already in-cluded in the analysis as one of the two influencing factors. How-ever, in the relationship between leader motivation and performance, where the two influencing factors are leader ability and the establishment of performance standards, the presence of a certain amount of member ability is of concern because it has not otherwise been included in the analysis.

Let me illustrate the point in the context of the basic training battalion example. Consider first the influence of member motivation on performance. When the members have high ability and the group is cohesive, the higher the members' motivation, the better the group will tend to perform. When the group has low ability or low cohesion, member motivation has little effect on performance. Thus, in determining what effect the motivation of Echo, Foxtrot, Golf, and Hotel have on performance, the members' ability to perform the task is taken into account. If that ability is low, as was the case with Golf, the members of that company simply do not have sufficient talent and skills to cause their high motivation to improve performance. When member ability is high, as was the case with Echo, Foxtrot, and Hotel, member motivation has a fairly strong influence on the outcome. The important point here is that the level of member ability was considered in the analysis.

When considering the effect of leader motivation on performance, however, the situation is different. In this relationship, the two influencing factors are the leader's ability and the establishment of performance standards. When the leader is competent and establishes high standards, his motivation has a fairly strong effect on performance. When the leader is incompetent or fails to establish high standards, his motivation has little effect on the outcome. Thus, in this relationship, the level of member ability is not taken into account, and we must determine whether or not the members possess at least a minimum level of skills to perform the task. If the members do possess the necessary ability, the rule or relationship, in this case the influence of leader motivation on performance, holds true. If the members do not possess the minimum level of skills necessary, which happens occasionally, the rule does not hold true. An example of the first case is Golf Company. Although Golf's cadre had below-average talent, these officers and noncommissioned officers clearly possessed enough skills to effectively use the good ideas and knowledge I offered them about better methods of training the new soldiers. Thus, Rule 9 held true and my high motivation had a positive effect on Golf's performance. In some instances, however, the members do not have the necessary minimum level of talent. For example, my first dining facility mess steward did not have the ability to effectively use the guidance and counseling he received. Even though I, as his leader, had above-average ability and had established high standards, my high motivation to have him perform well was not effective and, hence, Rule 9 did not hold true.

I must point out here, very emphatically, that the occasions when a rule does not hold true are both very infrequent and easily resolved. Normally, people are hired by an organization because they either possess certain skills or demonstrate the potential to acquire them. Further, people normally rise in an organization only to the level at which they can perform capably. Thus, it is clearly the exception to the norm for the leader to have a subordinate who has so little ability or potential that, even if that subordinate gives his best effort, he cannot be led or trained to do his job effectively. When those occasions do arise, as was the case with the mess steward, the problem is easily resolved by using that subordinate in another capacity for which he is better suited or by terminating his employment.

# 6
# Using the Rules for Leadership

The major benefit of these 10 rules for leadership is that they identify the conditions under which leadership style; enforcement of performance standards; member and leader intelligence, ability, and motivation; and group cohesion make positive contributions to group performance. Knowing these conditions allows the leader to determine which style will be most effective for him to use in his particular group situation. Also, this knowledge clearly points out which leader and group member qualities must be improved or maintained at a high level to achieve good group performance. The secondary benefit of this book is that it identifies a number of factors that serve to increase member and leader motivation and group cohesion.

Figure 2 displays the rules graphically, showing all 10 relationships at one time. The reader can thus determine which group qualities must be present in order for each of the 10 principal factors to have a positive influence on the outcome. For instance, Rule 1 shows that nondirective leadership combined with high member ability and high member motivation produces good unit performance.

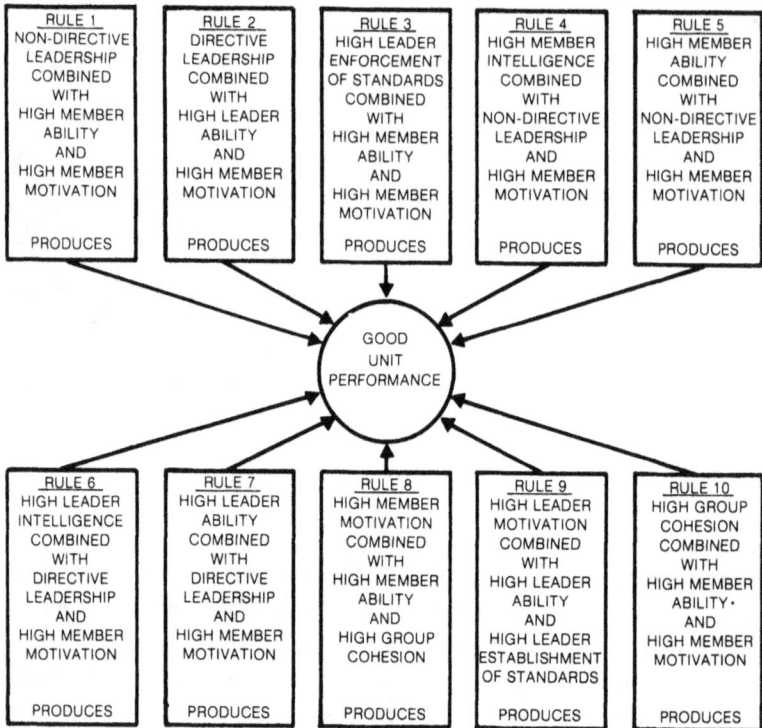

*Figure 2. Rules for Leadership*

## A Summary of the Rules

**Leadership Style (Rules 1 and 2).** The findings indicate that neither nondirective nor directive leadership by itself will increase the productivity of a group. Rather, the effectiveness of each leadership style depends upon the group's ability and the members' motivation to accomplish the task. Nondirective leadership, under which the members play the major role in planning and decisionmaking functions, is a good choice when one has talented subordinates. This style assures that high member ability contributes significantly to the task. When one has relatively unskilled or inexperienced subordinates, however, nondirective leadership is a poor choice. How can a style that depends predominantly on incompetent members result in high performance?

Directive leadership, which relies for the most part on the leader's expertise, is effective when the leader is talented. In this situation, the leader's good ideas and knowledgeable techniques will have a positive influence on the outcome. However, when the leader is inexperienced or doesn't know very much about his job, directive leadership is an inappropriate style for him to use. It just isn't very smart for the leader to tell people what to do and how to do it if his ideas aren't very good.

Even if both the members and the leader are very capable, it seems obvious that unless the members are motivated to accomplish the task, neither leadership style will be effective. If the members are not willing to use their own talents or to follow their leader's guidance, performance is going to be poor. It is no more useful to be nondirective with talented members who will not make a decent effort than it is to use this style with untalented members. Neither is it any more useful for the leader to be directive and provide excellent guidance to subordinates who will not follow his instructions than it is to provide poor advice.

When the members do have high motivation, the findings very clearly indicate which style the leader ought to use in any of the possible member and leader ability situations that he could encounter.

First, with high leader ability and low member ability, the directive style is the correct choice. The leader's good talents will have a positive influence on performance, while the members' lack of ability will have little effect.

Second, when the situation is reversed so that the members are talented and the leader is either inexperienced or just not very bright, nondirective leadership is the right choice. This style assures that the members' considerable skills, rather than the leader's low abilities, are the dominant influence on performance.

Third, in cases where neither the leader nor the members have very much ability, the unit is just not going to be able to perform well until something is done to improve ability. Regardless of whether the leader is directive and relies on his own poor skills or if he is nondirective and relies on the poor skills of his subordinates, performance is going to be low.

Fourth, in groups composed of a leader and members who both have high ability, either the nondirective or the directive style will be effective. The first will take advantage of high member skills, while the second will use the leader's good talents.

**Enforcement of Performance Standards (Rule 3).** The findings indicate that when the leader enforces high performance standards, which consists of both establishing a high minimum level of acceptable performance for one's subordinates and strongly insisting that those subordinates meet the established performance level, the group's productivity will be good, provided the members have both the ability and the motivation necessary to perform the task. With competent and enthusiastic subordinates, enforcement of high standards improves the outcome because it uses more of the members' skills and motivation. In this case, the leader causes the members to use all of their considerable ability and enthusiasm in order to reach the high standards he has set.

On the other hand, when the leader sets relatively low standards or does not insist that the members meet the high standards he has established, performance will be poor because the members need only use a portion of their capabilities and enthusiasm to reach a performance level the leader will accept. When the leader accepts whatever performance he gets from his subordinates, he usually won't get very much. In situations in which the members have low ability or poor motivation, the level to which the leader enforces standards has little effect on the outcome. It is difficult to see how the leader's attempts to enforce high standards can be effective if his subordinates just do not have the skills necessary to carry out his instructions, despite their best efforts. Further, it seems rather obvious that if the members are not willing to put forth a good effort, the outcome will be poor regardless of whether or not the leader tries to enforce high performance standards.

**Member and Leader Intelligence and Ability (Rules 4–7).** The findings indicate that the influence member and leader intelligence and ability have on group performance depends upon leadership style and member motivation. I distinguished between intelligence and ability by considering the first to be the talent used primarily during the planning and decisionmaking phase of the task and the second to be the specific skills and task-relevant knowledge required of people to actually accomplish the work.

Leadership style is important in these relationships because it determines whether the members' or the leader's talents are the dominant influence on performance. Under nondirective leadership, the members' skills play the major role; under directive leadership, the leader's skills have the most influence. Therefore, member intelligence and ability have a strong effect on the outcome when the leader uses the nondirective style. Of course, this style does not guarantee high performance. When the members are talented, performance will be good; when the members are relatively unskilled, performance will be poor. On the other hand, leader intelligence and ability have a strong effect on unit performance when the leader uses the directive style. When the leader is talented, performance is good, but when his ability is low, performance is poor.

The results also indicate that in situations in which the group members do not have sufficient motivation to accomplish the task, neither member nor leader intelligence or ability have any significant effect on the outcome. Because the members are the people who must actually perform the work and the leader cannot supervise in more than one place at a time, unless the members are willing to make a good effort, it matters little whether they are talented or not. Performance will be poor in either case. Similarly, if the members are unwilling to try wholeheartedly to carry out their leader's guidance, his talent is of little consequence.

**Member and Leader Motivation (Rules 8 and 9).** The findings indicate that member motivation has a strong influence on performance when the members have high ability and the group is cohesive. Certainly, when one's subordinates are capable of performing the task, the amount of effort they are willing to expend will greatly affect the outcome. When they try hard, performance will be good, but when they just "go through the motions" enough to stay out of trouble, the results will be poor. On the other hand, when the members have little talent, how can it matter very much whether or not they give the work their best effort? Even when they try hard, performance will be poor because they don't know what they are doing.

Group cohesion, which I defined as the extent to which the unit members display cooperative support, mutual assistance, organizational coordination, and teamwork in accomplishing the common task, must also affect the relationship between member motivation and group performance. In a cohesive group, the members use their group assets, including motivation, very efficiently. Thus, when these subordinates work hard, their efforts, which are effectively channeled toward improving the outcome, increase performance. On the other hand, if the members are poorly motivated, performance will be low despite the fact that the group gets along well. When the group is uncohesive, even if the members work hard, much of their effort is wasted in duplication of work, in gathering information that other members already have, and the like. Therefore, their motivation is spent unprofitably and adds little to the end result.

The results also indicate several factors that cause member motivation to increase. Among these are member perceptions that (1) their leader does his job well, (2) the members are given meaningful tasks to accomplish, (3) their leader has an interest in the members' opinions of how things should be done, (4) their leader is highly motivated, and (5) their leader enforces high performance standards. Because of questionnaire limitations, these factors are not meant to be interpreted as representing an all-inclusive list of subordinate motivators.

In the case of Rule 9, the findings indicate that the influence leader motivation has on performance depends upon the leader's ability and the level to which he establishes performance standards. Surely, when the leader is talented, the level of his motivation to contribute good ideas and knowledgeable techniques to his subordinates must affect the outcome. The competent leader who helps his subordinates by offering them suggestions, particularly when these group members are of below-average ability, will cause performance to improve. However, the talented leader who fails, through a lack of motivation, to contribute his advice to his subordinates does nothing to make the outcome any better. When the leader has low ability, it matters very little whether or not he is motivated to offer suggestions to the group members. How can offering poor advice be of any more value to the task than offering no advice at all?

The establishment of high standards also must affect the relationship between leader motivation and performance. When the leader sets high minimum levels of acceptable performance, the motivated leader will achieve excellent results because he insists that his subordinates meet those standards. The leader may have to require the members to repeat the task several times, but eventually they will come to understand that he will not accept shoddy performance and they will perform the task correctly. Conversely, when the leader is not motivated to insist that his subordinates perform the task to the high standards he has established, performance will most certainly be lower. In the situation where the leader establishes a low standard of performance, how can it matter whether or not he is motivated to ensure the group members meet that standard? Performance will be low and not appreciably different in either case.

The results also indicate that leader motivation has an indirect effect on the outcome by influencing other factors which themselves cause performance to be better or worse. Specifically, high leader motivation improves unit performance by raising the levels of leader enforcement of standards, member motivation, and group cohesion which, in their own right, positively influence the outcome.

The study identifies several factors that cause leader motivation to increase. Among these are high subordinate motivation and high leader job satisfaction. Other motivating factors include the leader's perceptions of whether his boss does his job well, asks for subordinates' opinions regarding how things should be done, is highly motivated, enforces high performance standards, and assigns worthwhile and important tasks to subordinates.

**Group Cohesion (Rule 10).** The findings indicate that when the group has high cohesion, performance will be good, provided the members have both the ability and the motivation necessary to accomplish the task. With competent and enthusiastic subordinates, high unit cohesion improves the outcome because the members' good working relationship uses group assets such as ability, motivation, and time more efficiently and effectively. When the members pool their ideas and knowledge, divide up the common task among themselves, and the like, performance is going to be much better than when the group is uncohesive and does not do these things. In the second case, much of the members' talents and efforts is wasted in performing activities that do little to improve the outcome, such as duplicating effort and using less-efficient work techniques when other group members know better methods.

Regarding member ability, it seems to be necessary for one's subordinates to have a certain amount of skill in order for cohesion to improve performance. In other words, if the members have little talent, the outcome will be relatively poor regardless of whether or not they are cohesive and work well together. It is hard to see how a cooperative effort among people who have little ability can result in good performance. Further, unless one's subordinates are willing to make a good effort to accomplish the task, it matters little whether the group is cohesive or not. Regardless of the good working relationship the group members may have among themselves, if they do only the minimum amount of work necessary to stay out of trouble, the outcome will be poor.

The results also indicate several factors that cause group cohesion to improve. Among these are member perceptions that the leader enforces high performance standards, does his job well, is motivated to accomplish the task, and asks his subordinates for their opinions concerning the work.

### Recommendations for Using the Rules

**Evaluation.** The first step the leader ought to take to use these rules is to evaluate himself and his subordinates in terms of the principal and influencing factors discussed in this book. In other words, he should sit down and rate himself and the people who work for him as being high or low on the relevant factors:

|     *Leader*              |     *Subordinates*   |
| ------------------------- | -------------------- |
| 1. Enforcement of         | 1. Intelligence      |
|    Standards              | 2. Ability           |
| 2. Intelligence           | 3. Motivation        |
| 3. Ability                | 4. Group Cohesion    |
| 4. Motivation             |                      |
| 5. Group Cohesion         |                      |

It is critical that the leader's evaluations be as accurate as possible because these ratings form the basis upon which he will make his decisions about which leadership style to use and which leader and group member qualities must be improved.

**Member Motivation.** The next logical step is to focus on how much effort one's subordinates are willing to make to accomplish the task. High motivation is probably the most critical quality for the members to have because it is directly involved as a principal or influencing factor in 9 of the 10 rules. As a principal factor, high member motivation is important because it causes considerably better performance than does low member motivation. Simply put, if the members are talented and the group is cohesive, when one's subordinates try hard, they put into productive use a greater amount of their ability than when they do not try hard. High member motivation is equally important as an influencing factor because it determines whether or not leadership style, enforcement of high performance standards, high member and leader intelligence and ability, and high group cohesion will positively influence the outcome. When member motivation is low, none of these factors cause performance to improve. Further, although the leader can "make up for" low member intelligence and ability by using the directive leadership style, he cannot make up for low member motivation by trying harder himself. After all, he cannot physically do the work of 10 of his subordinates nor can he supervise the group members in more than one place at a time. Thus, when one's subordinates have low motivation, the leader's first priority should be to correct this problem.

A good first step to take in order to improve member motivation is to sit down with the group members and find out what problems are bothering them or what changes and incentives would cause them to work harder. In the majority of cases, the leader can usually resolve "people problems" fairly easily by taking an active interest. The simple act of sitting down with one's subordinates and talking things over with them serves as a motivator by itself. Some of the other factors that motivate group members are presented earlier in this chapter and in Chapter 4. The leader should not take the view that he is "giving in" to his subordinates or that they are "putting one over" on him if he responds favorably to a legitimate complaint or reasonable request by the group members. It is the right, not a privilege, of subordinates to be treated well and fairly. The leader who takes this view will go a long way toward gaining the good member support for the task that is so essential for high unit performance.

**Member and Leader Intelligence and Ability.** The leader's next concern should be the talent and skills of himself and his subordinates. After all, although there is no point in worrying about someone's ability if he does not have the motivation to use it, there is also no point in worrying about someone's motivation unless he has the skills necessary to perform the task. Every one of the 10 rules is influenced, in one way or another, by either member or leader intelligence or ability. Intelligence is involved only as a principal factor. In that role, high intelligence is important because it causes better performance than does low intelligence. Specifically, high intelligence has a positive influence on the outcome because more intelligent people can contribute better ideas to the planning and make better decisions regarding alternative choices than can less intelligent people, and better planning and decision-making result in better performance.

Member and leader ability, which influence primarily the performance phase, serve both as principal and influencing factors. In their role as principal factors, high member and leader ability cause better unit performance than does low ability because people with more talent are capable of doing better at whatever skill the task requires. The difference between good and bad performance is, of course, no more than the difference in how much ability the group displays. As influencing factors, member and leader ability are important because they determine whether or not other factors will positively influence the outcome. When member ability is low, nondirective leadership, enforcement of high performance standards, high member motivation, and high group cohesion do not cause performance to improve. When leader ability is low, neither directive leadership nor high leader motivation will improve the outcome.

The solution to low ability problems is, of course, training. In the short run, the bright leader can "cover up" for his subordinates' lack of ability by using the directive style, and the inexperienced or incompetent leader can "get away with" using the nondirective style if he has talented group members. But, when the bright leader or the talented members are lost to the group, the unit is in a very "iffy" situation. Unless equally competent replacements are available, which may be unlikely, the unit is in trouble. The dependable solution to low ability problems is training. Whether the leader chooses to use away-from-unit schooling for some job specialties, in-unit classes and field training, on-the-job experience, or a combination of these methods, the rules clearly indicate that efforts spent in training the unit members are well worthwhile in terms of improving unit performance.

**Leadership Style.** The choice of which leadership style to use is obviously very important to unit performance. The reason I discuss this topic after, rather than before, member motivation and member and leader intelligence and ability is that one cannot determine which styles will be effective for him until he has determined whether the members are enthusiastic or not and what level of talent or ability he and his subordinates possess. Once the leader resolves these two concerns, he can decide whether to be directive, nondirective, or some of each. Moreover, as explained in my basic training battalion example, a leader can choose to be directive with some units or members and nondirective with others. Regardless of the style the leader uses, it is important that he clearly indicates to his subordinates where the planning, decisionmaking, and supervisory responsibilities lie.

The basic considerations in choosing a leadership style are member motivation and member and leader ability. Regarding member motivation, neither leadership style will be effective when one's subordinates are not willing to try hard. It makes no difference whether the leader tells the group members what to do and how to do it or if he gives them the authority to make their own decisions; if the group members do not care enough to make a good effort, performance is going to be poor.

In terms of ability, if the members have high motivation, and they and the leader are both talented, either the directive or the nondirective style will be effective. Also, in this case, the leader can choose to be equally directive and nondirective. In other words, he can retain decisionmaking control for certain group tasks or functions and in others, allow the members to perform the work as they see fit. If the members have low ability and the leader is talented, he should use the directive style. To be nondirective and allow incompetent members to perform the task as they see fit is to court disaster. If the members are talented and the leader inexperienced or just not very knowledgeable regarding the task, he ought to use the nondirective style. When this type of leader is directive and requires his subordinates to do things his way, the result will be poor performance. If neither the leader nor the members are very talented, the leader had better start training his subordinates and "burning the midnight oil" for himself. If he doesn't, he's going to lose his job in fairly short order. Most commanders will solidly back a subordinate who takes over a poor unit as long as they feel the subordinate has the talent necessary to make the unit better and they see some signs of unit improvement. But when a commander feels that one of his subordinate leaders does not have the talent to cause his unit to perform well or, at least, make progress, the subordinate leader's days in that unit are numbered.

**Enforcement of Performance Standards.** Although the remaining three factors, enforcement of performance standards, leader motivation, and group cohesion, can each significantly improve unit performance, they do not have quite the importance of member motivation, member and leader intelligence and ability, and leadership style. The reason for the difference is that the first three are each involved in only one or two rules, while the second three are each involved in the majority of rules and, thus, have a more widespread effect.

In considering enforcement of performance standards, leader motivation, and group cohesion, there is very little reason to attend to one of these factors before the others. Each of the factors is important, and the leader should make sure he gives them the attention they deserve. Enforcement of high standards, of course, consists of the leader establishing a high minimum level of acceptable performance for his group members and strongly insisting that his subordinates meet that established level. As a principal factor, enforcement of high standards causes better unit performance than does the failure to enforce high standards because the first uses a greater amount of the group members' ability and motivation, provided they are competent and enthusiastic. As an influencing factor, enforcement of standards is important because it determines whether or not high leader motivation positively influences the outcome. When the leader establishes low standards, his high motivation is not able to improve performance.

Unfortunately, too many leaders will not say, "That's not good enough, do it again," or the equivalent. One reason is low leader ability. If a leader is inept, what basis does he have for judging the performance of his subordinates? None, of course. It is a necessity, then, if the leader intends to enforce high performance standards, that he learns how the task should be done correctly. The leader who "inspects training" and doesn't know what to look for is just wasting his time.

A second reason for failure to enforce high standards is low leader motivation. If the leader simply does not care how his people perform the task or if he is unwilling to inspect their work when it is raining, cold, hot, or in the middle of the night, he cannot possibly enforce high standards. Enforcing standards requires direct observation. In my basic training battalion example, when Echo fired a company average rifle marksmanship score of 28 instead of its normal 31 on one particular December day, I'd have looked like a fool if I'd made them practice some more without having inspected their initial firing. But, because I'd watched them shoot in the driving snow, sleet, and freezing rain, I knew their score of 28 was excellent under the circumstances and they needed no more practice.

A third reason for failure to enforce high standards is that the leader is more interested in being "popular" than he is in accomplishing the unit mission. If, after a unit spends a hard day conducting a combat operation in a steaming jungle, the leader requires his "dog-tired" soldiers to dig foxholes for the night's encampment, he will not be very "popular." But, the next morning, soldiers who might otherwise be dead will have a "healthy" respect for their leader and will be able to perform that day's mission. The leader who unnecessarily sends soldiers home in "body bags" because of his failure to "insist upon the harder right, rather than going along with the easier wrong" won't be "popular" for very long.

**Leader Motivation.** The level to which the leader is motivated to accomplish the task is important for two reasons. First, if the leader has high ability and establishes high standards, the highly motivated leader causes better unit performance than does the poorly motivated leader because the former contributes more good ideas and suggestions to the task than does the latter, and the former requires greater effort, both quantitatively and qualitatively, from the group members during the performance phase. The second reason high leader motivation is important is that it raises the levels of leader enforcement of standards, member motivation, and group cohesion which, by themselves, have a positive effect on the outcome.

The group leader, for example the company commander, is not the only person we should be concerned about when we discuss ways to raise low leader motivation. We should also be concerned with the unit's subordinate leaders and with the leader's boss because the motivation of these people will clearly have an effect on the company's performance. The motivation of the subordinate (platoon) leaders is important because their willingness to make a good effort has a direct influence on how well the individual platoons perform. When these subordinate leaders try hard, their units will do well, and the quality of company performance is, in most cases, just the sum of how well each of the platoons does its job. The motivation of the company commander's boss, the battalion commander, affects company performance in two ways. First, if the battalion commander is willing to make suggestions about better techniques and work methods for accomplishing the task, he will be of great assistance to the company. Second, if the battalion commander is motivated to insist upon high performance from each company, he provides strong support for the company commander's efforts to get good performance from his platoons. With the "clout" of the battalion commander to back him up, the company commander has more perceived authority, which aids his efforts to enforce high standards. The importance of the company commander's own motivation to unit performance is very similar to the effect of the battalion commander's motivation. When the company commander is motivated to give advice to his platoons and insists that the platoons meet high performance standards, the company will do well.

The factors that motivate the different leaders discussed in the preceding paragraph are virtually the same for each of them. Some of these motivating factors include the perceptions of whether one's boss (1) does his job well, (2) asks his subordinate leaders for their opinions about how things should be done, (3) is motivated to accomplish the task, (4) enforces high performance standards, (5) assigns worthwhile and important tasks to his subordinate leaders, and (6) treats his subordinate leaders decently and fairly. These factors are some of the things that commanders at every level should do to improve their subordinate leaders' motivation.

**Group Cohesion.** The level of group cohesion is important for two reasons. As a principal factor, if the group members are competent and enthusiastic, high group cohesion will cause better unit performance than will low group cohesion because the former more efficiently and effectively uses group assets such as ability, motivation, and time. As an influencing factor, group cohesion is important because it determines whether or not high member motivation will positively influence the outcome. When group cohesion is low, high member motivation has considerably less effect on performance.

Several things can be done to improve group cohesion. Some of these actions are within the unit leader's capability, and others are the responsibility of the leaders at the highest levels of the parent organization. Within the US Army, examples of high-level actions include (1) recruiting "Cohort" platoons, whereby soldiers from the same hometown or area are enlisted and assigned to serve in the same platoon for their full first term of service, and (2) the "home basing" or "regimental system," under which a career soldier serves the majority of his assignments within the various battalions of a single regiment. The rationale for Cohort platoons is that unit cohesion will improve because soldiers who grow up in the same area are more likely to share similar values, interests, and attitudes and to be closer friends. The rationale for the regimental system is that soldiers who serve multiple tours of duty within the same regiment will develop a strong allegiance to that unit and will establish close bonds with other unit members by serving with them on repeated occasions.

One of the actions a unit leader can take to improve the cohesion within his own group is to foster pride in unit membership. In the basic training battalion, it was important for me to convince the company commanders and their drill sergeants that they were members of the best battalion on the post and that their own particular company was the best company within the battalion. The leader can help to develop these perceptions by (1) initially assigning tasks that his subordinates will be able to perform well, (2) strongly emphasizing excellent appearance of the unit area and the soldiers themselves, (3) encouraging the companies to have unit mottoes, signs, and distinctive apparel with the company symbol and slogan printed on them, and (4) ensuring that the companies are rewarded with trophies, streamers, letters of commendation, or the like for their successes.

The leader can also improve group cohesion by strongly establishing the worth of the parent organization and its values. Subordinates will develop a sense of commitment to an organization they feel is a "winner." For example, the leader's enthusiasm and supporting words for the Army's proud heritage, traditions, and record of accomplishments pay dividends in unit cohesion. When a leader "badmouths" the Army, he hurts his own unit. Other factors that develop high unit cohesion are subordinate perceptions that the leader enforces high performance standards, does his job well, is motivated to accomplish the unit mission, asks his subordinates for their opinions concerning the work, and treats the unit members with fairness and equity.

### Conclusion

The information in this book has important implications for both practical leadership and theory. For leaders in the field, I have tried to demonstrate several things. First, one's decision regarding which leadership style to use should be based upon the levels of his own ability and the ability and motivation of his subordinates. To choose a style based on what worked for someone else, or upon other factors, is to invite disaster. Second, enforcement of high performance standards; high member and leader intelligence, ability, and motivation; and high group cohesion are worth working hard to achieve because these factors have the potential to considerably improve group performance. One must remember, however, that whether or not this potential will be realized depends, as explained in each of the 10 rules, on whether certain other factors are present. For example, it will do a leader little good to devote great effort to improving his subordinates' abilities if he ignores the level of their motivation. Third, there are some very specific things one can do to improve member and leader ability and motivation, group cohesion, and the leader's ability to enforce performance standards. These actions, summarized in the preceding section of this chapter, have proven to be effective in actual Army units and are very much worth consideration.

For theorists, the most significant findings concern the effects that leadership style; enforcement of performance standards; member and leader intelligence, ability, and motivation; and group cohesion have on unit performance. The results show that each of these factors can have a fairly strong effect on the unit's productivity and that the actual amount of influence these factors make in any given case depends upon the presence or absence of certain other specific factors. These findings not only tend to clarify the confusing picture in the research literature but also clarify the processes that determine group or unit effectiveness.

Finally, I should briefly discuss the issue of study generalizability, the degree that the findings of this study will be applicable in other settings. It is certainly important whether a sufficient basis exists for the reader to feel that the findings of this book are applicable to the group situation in which he presently finds himself or anticipates being in at a future time. One of the book's primary strengths is that the findings are based on data collected during a field study of 49 actual groups or units, rather than on data collected during a laboratory experiment. The subjects were permanent members of existing groups, and the tasks upon which they were evaluated were their normal everyday duties. Although, in a field study, there are somewhat more outside distractions, such as time constraints on unit members to complete study tests and questionnaires and the inability to collect data from unit members who were on military leave or otherwise unavailable, the value of having opinions on group performance and individual behaviors from people who have worked together and observed one another for many months is a very strong and credible asset that laboratory studies almost never have. The fact that the 10 rules in this book did hold true in everyday "real-world" groups is a strong "plus" for the generalizability of the findings.

Another aspect of this study that affects its generalizability is that the subject groups were military units. Within the military, certain organizational characteristics and operating procedures are fairly consistent from unit to unit. For example, the relationship between the leader and his subordinates is well-defined and tends to be fairly formal. Also, the average leadership style tends more toward the directive approach, and the work tends to be fairly structured or, in other words, well-understood and done according to certain prescribed methods. Thus, there is a "plus" for the applicability of this study's findings to other military units and to groups that have similar characteristics, while there may or may not be a "minus" with regard to organizations which have opposite characteristics of leader-member authority, leadership style, and task structure.

Although some may feel that the job requirements of the leaders (mess stewards) in this study are fairly undemanding and, therefore, detract from the generalizability of the study to more complex leadership roles, it is my opinion that it is no easy task to manage an Army dining facility successfully. For example, the dining facility of the basic training battalion to which I was assigned was required to prepare and serve, on the average, 3,000 meals a day, seven days a week. The mess steward was responsible for preparing each menu; procuring and properly storing the food; preparing and serving the meals; ensuring the cleanliness of the dining facility, utensils, and equipment; delivering and serving meals at the correct time at multiple training sites; estimating how much of each type of food should be prepared for each meal; and training the 18–27 cooks who worked for him, of whom over 70 percent were serving in their first assignment as cooks. Most important, the mess steward had fiscal responsibility for the entire dining facility operation. He had a three-month operating budget with which to purchase the necessary food and supplies and, after serving some 270,000 meals during this period, was required to be within 3 percent of his initially allocated funds. Thus, I do not believe the mess steward's job requirements can be described as simple or undemanding, and, therefore, the study findings do not lose their generalizability to complex leadership positions. Additional factors can be presented which further support this study's generalizability to other settings. I believe, however, that the three factors just presented make the point sufficiently.

The saying, "Leaders are not born, they are made," is true. I earnestly hope that those in leadership positions will use these "rules for leadership" because I firmly believe that the rules will "make" good leaders and improve unit performance.

# Bibliography

Aronson, E., and Carlsmith, J. M. "Performance Expectancy as a Determinant of Actual Performance." *Journal of Abnormal and Social Psychology,* 1962, *65,* 178–182.

Atkinson, J. W. "Towards Experimental Analysis of Human Motivation in Terms of Motives, Expectancies, and Incentives." In J. W. Atkinson, ed. *Motives in Fantasy, Action and Society.* Princeton: Van Nostrand, 1958, 288–305.

———— and Reitman, W. R. "Performance as a Function of Motive Strength and Expectancy of Goal Attainment." *Journal of Abnormal and Social Psychology,* 1956, *53,* 361–366.

Bavelas, A. "Communications Patterns in Task-Oriented Groups." *Journal of the Acoustical Society of America,* 1950, *22,* 725–730.

Berkowitz, L. "Group Norms Among Bomber Crews: Patterns of Perceived Crew Attitudes, 'Actual' Crew Attitudes, and Crew Liking Related to Air-Crew Effectiveness in Far Eastern Combat." *Sociometry,* 1956, *19,* 141–153.

Bjerstedt, A. "Preparation, Process, and Product in Small Group Interaction." *Human Relations,* 1961, *14,* 183–189.

Blades, J. W. "The Influence of Intelligence, Task Ability, and Motivation on Group Performance." Unpublished Ph.D. dissertation, University of Washington, 1976.

Blake, R. R. and Mouton, J. H. *The Managerial Grid.* Houston: Gulf Publishing Co., 1964.

Bridgeman, W. "Student Attraction and Productivity as a Composite Function of Reinforcement and Expectancy Conditions." *Journal of Personality and Social Psychology*, 1972, *23*, 249–258.

Buros, O. D., ed. *The Sixth Mental Measurements Yearbook*. Highland Park, N. J.: Gryphon Press, 1965.

Calvin, A. D.; Hoffman, T. K.; and Harden, E. C. "The Effect of Intelligence and Social Atmosphere on Group Problem Solving Behavior." *Journal of Social Psychology*, 1957, *45*, 61–74.

Campbell, J. P.; Dunnette, M.D.; Lawler, E. E.; and Weick, K. E. *Managerial Behavior, Performance, and Effectiveness*. New York: McGraw-Hill, 1970.

Campion, J. E. "Effects of Managerial Style on Subordinates' Attitudes and Performance in a Simulated Organization Setting." Unpublished Ph.D. dissertation, University of Minnesota, 1968.

Cartwright, D., and Zander, A. F., eds. *Group Dynamics: Research and Theory*. 2nd ed. Evanston, Ill.: Row, Peterson, 1960.

Castaneda, A., and Palermo, D. S. "Psychomotor Performance as a Function of Amount of Training and Stress." *Journal of Experimental Psychology*, 1955, *50*, 175–179.

Coch, L., and French, J. R. P. "Overcoming Resistance to Change." *Human Relations*, 1948, *1*, 512–532.

Collins, B. E., and Raven, B. H. "Group Structure: Attraction, Coalitions, Communications, and Power." In G. Lindzey and E. Aronson, eds. *The Handbook of Social Psychology*, Vol. IV. Reading, Mass.: Addison-Wesley, 1969.

Comrey, A. L. "Group Performance in a Manual Dexterity Task." *Journal of Applied Psychology*, 1953, *37*, 207–210.

Csoka, L. S. "A Contingency Model Approach to Leadership Training and Leadership Experience." Unpublished M.S. thesis, University of Washington, 1971.

Davis, K. *Human Relations at Work*. New York: McGraw-Hill, 1962.

Drucker, P. F. *The Practice of Management*. New York: Harper, 1954.

Fiedler, F. E. *A Theory of Leadership Effectiveness*. New York: McGraw-Hill. 1967.

Fiedler, F. E., and Meuwese, W. A. T. "The Leader's Contribution to Task Performance in Cohesive and Uncohesive Groups." *Journal of Abnormal and Social Psychology,* 1963, *67,* 83–87.

Fleishman, E. A. "A Relationship between Incentive Motivation and Ability Level in Psychomotor Performance." *Journal of Experimental Psychology,* 1958, *56,* 78–81.

_____. *Manual for Leadership Opinion Questionnaires.* Chicago: Science Research Associates, 1960.

_____; Harris, E. F.; and Burtt, H.E. *Leadership and Supervision in Industry.* Columbus, Ohio: Ohio State University, 1955.

French, E. G. "Effects of Interaction of Achievement, Motivation, and Intelligence on Problem Solving Success." *American Psychologist,* 1957, *12,* 399–400 (Abstract).

French, J. R.; Israel, P. J.; and As, D. "An Experiment on Participation in a Norwegian Factory." *Human Relations,* 1960, *13,* 3–19.

Gagne, R. M., and Fleishman, E. A. *Psychology and Human Performance.* New York: Holt, 1959.

Georgopoulus, B. S.; Mahoney, G. M.; and Jones, N. W. "A Path-Goal Approach to Productivity." *Journal of Applied Psychology,* 1957, *41,* 345–353.

Ghiselli, E. E. "The Validity of a Personnel Interview." *Personnel Psychology,* 1966, *19,* 389-395.

Greer, F. T.; Galanter, E. H.; and Nordlie, P. G. "Interpersonal Knowledge and Individual and Group Effectiveness." *Journal of Abnormal and Social Psychology,* 1954, *49,* 411–414.

Halpin, A. W., and Winer, B. J. "A Factorial Study of Leader Behavior Descriptions." In R. M. Stogdill and A. E. Coons, eds. *Leader Behavior: Its Description and Measurement.* Columbus, Ohio: Ohio State University, Bureau of Business Research, Monograph No. 88, 1957, 39–51.

Havron, M. D., and McGrath, J. E. "The Contribution of the Leader to the Effectiveness of Small Military Groups." In L. Petrullo and B. M. Bass, eds. *Leadership and Interpersonal Behavior.* New York: Holt, Rinehart, and Winston, 1952.

Hersey, P., and Blanchard, K. H. *Management of Organizational Behavior: Utilizing Human Resources*. Englewood Cliffs, New Jersey: Prentice-Hall, 1969.

Heslin, R. "Predicting Group Task Effectiveness from Member Characteristics." *Psychological Bulletin*, 1964, *62*, 248–256.

Hoffman, L. R. "Homogeneity of Member Personality and its Effect on Group Problem Solving." *Journal of Abnormal and Social Psychology*, 1959, *58*, 27–32.

Homans, G. C. *The Human Group*. New York: Harcourt Brace, 1950.

――――. "Group Factors in Worker Productivity." In E. E. Maccoby, T. M. Newcomb, and E. L. Hartley, eds. *Readings in Social Psychology*. New York: Holt, Rinehart, and Winston, 1958.

Horsfall, A. B., and Arensberg, C. M. "Teamwork and Productivity in a Shoe Factory." *Human Organizations*, 1949, *8*, 13–25.

Johns, J. H., ed. *Cohesion in the US Military*. Washington, DC: National Defense University Press, 1984.

Katz, D.; Maccoby, N.; Gurin, G.; and Floor, L. G. *Productivity, Supervision, and Morale Among Railroad Workers*. Ann Arbor, Michigan: University of Michigan Institute for Social Research, 1951.

――――.; Maccoby, N.; and Morse, N.C. *Productivity, Supervision, and Morale in an Office Situation*. Ann Arbor, Michigan: University of Michigan Institute for Social Research, 1950.

Kaufman, H. "Task Performance, Expected Performance, and Responses to Failure as Functions of Imbalance in the Self-Concept." Unpublished Ph.D. dissertation, University cf Pennsylvania, 1962.

Korman, A. K. "The Prediction of Managerial Performance: A Review." *Personnel Psychology*, 1968, *21*, 295–322.

Lange, C. J.; Campbell, V.; Katter, R.V.; and Shanley, F. J., *A Study of Leadership in Army Infantry Platoons*. Research Report,. Monterey, California: US Army Human Research Unit, 1958.

Likert, R. "A Motivational Approach to a Modified Theory of Organization and Management." In M. Haire, ed. *Modern Organizational Theory: A Symposium of the Foundation for Research on Human Behavior*. New York: Wiley, 1959, 184–217.

————. *The Human Organization.* New York: McGraw-Hill, 1967.

Maier, N. R. F. *Psychology in Industry.* 2nd ed. Boston: Houghton-Mifflin, 1955.

————. *Problem-Solving Discussions and Conferences.* New York: McGraw-Hill, 1963.

Mann, R. D. "A Review of the Relationships between Personality and Performance in Small Groups." *Psychological Bulletin,* 1959, *56,* 241–270.

Marquis, D. G.; Guetzkow, H.; and Heyns, R. W. "A Social Psychological Study of the Decision-Making Conference." In H. Guetzkow, ed. *Groups, Leadership and Men: Research in Human Relations.* Pittsburgh: Carnegie Press, 1951.

McGrath, J. E., and Altman, J. *Small Group Research.* New York: Holt, Rinehart, and Winston, 1966.

McGregor, D. *Leadership and Motivation.* Cambridge, Mass: MIT Press, 1966.

McKeachie, W. J. "Student-Centered versus Instructor-Centered Instruction." *Journal of Educational Psychology,* 1954, *45,* 143–150.

McNamara, V. D. "A Descriptive—Analytic Study of Directive—Permissive Variation in the Leader Behavior of Elementary School Principals." Unpublished M.S. thesis, University of Alberta, 1967.

Meuwese, W. A. T., and Fiedler, F. E. *Leadership and Group Creativity under Varying Conditions of Stress.* Technical Report. Urbana, Ill.: University of Illinois Group Effectiveness Research Laboratory, 1965.

Middlebrook, P. N. *Social Psychology and Modern Life.* New York: Knopf, 1974.

Morse, N. C., and Reimer, E. "The Experimental Change of a Major Organizational Variable." *Journal of Abnormal and Social Psychology,* 1956, *52,* 120–129.

O'Brien, G. E., and Owens, A. G. *Effects of Organizational Structure upon Correlations between Member Abilities and Group Productivity.* Technical Report. Urbana, Ill.: University of Illinois Group Effectiveness Research Laboratory, 1969.

Pelz, D. C. "Some Social Factors Related to Performance in a Research Organization." *Administrative Science Quarterly*, 1956, *1*, 310–325.

Peters, T. J., and Waterman, R. H., Jr. *In Search of Excellence*. New York: Harper and Row, 1982.

Posthuma, A. B. *Normative Data on the Least Preferred Co-worker (LPC) Scale and the Group Atmosphere (GA) Scale*. Technical Report. Seattle: University of Washington Organizational Research Group, 1970.

Rambo, W. W. "The Construction and Analysis of a Leadership Behavior Rating Form." *Journal of Applied Psychology*, 1958, *42*, 409–415.

Rhode, K. J. "Theoretical and Experimental Analysis of Leadership Ability." *Psychological Reports*, 1958, *4*, 243–278.

Schachter, S.; Ellertson, N.; McBride, D.; and Gregory, D. "An Experimental Study of Cohesiveness and Productivity." *Human Relations*, 1951, *4*, 229–238.

Schultz, W. C. *FIRO: A Three-Dimensional Theory of Interpersonal Behavior*. New York: Rinehart, 1958.

Shure, G. H.; Rogers, M.S.; Larson, J. M.; and Tassone, J. "Group Planning and Task Effectiveness." *Sociometry*, 1962, *25*, 263–282.

Spector, P., and Suttell, B. J. *An Experimental Comparison of the Effectiveness of Three Patterns of Leadership Behavior*. Technical Report. Washington, DC: American Institute for Research, 1957.

Stogdill, R. M. *Manual for the Leader Behavior Description Questionnaire-Form XII: An Experimental Revision*. Columbus, Ohio: Ohio State University, 1963.

_____. *Handbook of Leadership*. New York: The Free Press, 1974.

Sun Tzu. *Sun Tzu: The Art of War*. Translated by S. B. Griffin. London: Oxford University Press, 1963.

Thoms, E. J., and Fink, C. F. "Models of Group Problem Solving." *Journal of Abnormal and Social Psychology*, 1961, *63*, 53–63.

US, Department of Defense. *Project One Hundred Thousand*. Washington, DC: US Government Printing Office, 1969.

Van Zelst, R. "Validation of a Sociometric Regrouping Procedure." *Journal of Abnormal and Social Psychology,* 1952, *47,* 299–301.

Vroom, V. H. "Some Personality Determinants of the Effects of Participation." *Journal of Abnormal and Social Psychology,* 1959, *59,* 322–327.

———. "Ego-Involvement, Job Satisfaction, and Job Performance." *Personnel Psychology,* 1962, *15,* 159–177.

———. *Work and Motivation.* New York: John Wiley and Sons, Inc., 1964.

Wyatt, S. *Incentives in Repetitive Work: A Practical Experiment in a Factory.* Research Report No. 69. London: H. M. Stationery Office, 1937.

# The Author

Lieutenant Colonel Jon W. Blades, US Army, wrote this book while attending the National War College where he was also a Senior Fellow at the National Defense University. The author has commanded infantry companies in Germany and Vietnam; served as a brigade/battalion operations officer in Germany, Vietnam, and Korea; and commanded a basic training battalion at Fort Jackson, South Carolina. Lieutenant Colonel Blades holds a Ph.D. in Social Psychology (Organizational Research). He has taught military psychology and leadership at West Point and has served with the Leadership Division, Office of the Deputy Chief of Staff for Personnel, Department of the Army.